The Cerebral Circulation

Integrated System Physiology: From Molecule to Function

Editors

D. Neil Granger, *Louisiana State University Health Sciences Center*
Joey Granger, *University of Mississippi Medical Center*

This Series focuses on combining the full spectrum of tissue function from the molecular to the cellular level. Topics are treated in full recognition that molecular reductionism and integrated functional overviews need to be combined for a full view of the mechanisms and biomedical implications.

Published Titles:

The Cerebral Circulation
Marilyn J. Cipolla
University of Vermont College of Medicine

Hepatic Circulation
W. Wayne Lautt
University of Manitoba

Forthcoming Titles:

Capillary Fluid Exchange
Ronald Korthuis
University of Missoui, Columbia

Cutaneous Circulation
Nicholas Flavahan
Johns Hopkins University

Endothelin and Cardiovascular Regulation
David Webb
University of Edinburgh, Queen's Medical Institute

Homeostasis and the Vascular Wall
Rolando Rumbaut
Baylor College of Medicine

Inflammation and Circulation
D. Neil Granger
Louisiana State University Health Sciences Center

Lymphatics
David Zawieja
Texas A&M University

Pulmonary Circulation
Mary Townsley
University of South Alabama

Regulation of Arterial Pressure
Joey Granger
University of Mississippi

Regulation Cardiac Output
David Young
University of Mississippi Medical School

Regulation of Tissue Oxygenation
Roland Pittman
Virginia Commonwealth University

Regulation of Vascular Smooth Muscle Function
Raouf Khalil
Harvard University

Vascular Biology of the Placenta
Yuping Wang
Louisiana State University

The Cerebral Circulation
Marilyn J. Cipolla
www.morganclaypool.com

ISBN: 9781615040124   paperback

ISBN: 9781615040131   ebook

DOI: 10.4199/C00005ED1V01Y200912ISP002

A Publication in the Morgan & Claypool Life Sciences series

INTEGRATED SYSTEMS PHYSIOLOGY: FROM MOLECULE TO FUNCTION

Book #2

Series Editors: D. Neil Granger, Louisiana State University; Joey Granger, University of Mississippi

Series ISSN Pending

The Cerebral Circulation

Marilyn J. Cipolla
University of Vermont College of Medicine

INTEGRATED SYSTEMS PHYSIOLOGY: FROM MOLECULE TO FUNCTION #2

ABSTRACT

This presentation describes structural and functional properties of the cerebral circulation that are unique to the brain, an organ with high metabolic demands, and the need for tight water and ion homeostasis. Autoregulation is pronounced in the brain, with myogenic, metabolic, and neurogenic mechanisms contributing to maintain relatively constant blood flow during both increases and decreases in pressure. In addition, unlike peripheral organs where the majority of vascular resistance resides in small arteries and arterioles, large extracranial and intracranial arteries contribute significantly to vascular resistance in the brain. The prominent role of large arteries in cerebrovascular resistance helps maintain blood flow and protect downstream vessels during changes in perfusion pressure. The cerebral endothelium is also unique in that its barrier properties are in some way more like epithelium than endothelium in the periphery. The cerebral endothelium, known as the blood–brain barrier, has specialized tight junctions that do not allow ions to pass freely and has very low hydraulic conductivity and transcellular transport. This special configuration modifies Starling's forces in the brain such that ions retained in the vascular lumen oppose water movement due to hydrostatic pressure. Tight water regulation is necessary in the brain because it has limited capacity for expansion within the skull. Increased intracranial pressure due to vasogenic edema can cause severe neurologic complications and death. This chapter will review these special features of the cerebral circulation and how they contribute to the physiology of the brain.

KEYWORDS

cerebral circulation, neurovascular unit, blood–brain barrier, myogenic, autoregulation

Contents

CHAPTER 1

Introduction

As an organ, the brain comprises only about 2% of body weight yet it receives 15–20% of total cardiac output, making the brain one of the most highly perfused organs in the body. The high metabolic needs of the brain, relying heavily on oxidative metabolism, necessitate not only a high fraction of cardiac output but also relatively constant blood flow. The brain is also unique in that it is enclosed by the skull, a bony rigid structure that does not allow for expansion of either tissue or extracellular fluid without significant deleterious effects. Swelling of the brain due to vasogenic edema can increase intracranial pressure (ICP) and cause severe neurologic complications and even death. Because of the importance to maintain ICP within normal ranges and also to provide an appropriate ionic milieu for neuronal function, water and solute transport from the blood into the brain parenchyma is controlled in very special ways. The cerebral circulation is also unique in that the large arteries account for a greater proportion of vascular resistance in the brain than in many other vascular beds. This unusually prominent role of large arteries in vascular resistance likely helps to provide constant blood flow to neuronal tissue and protect the cerebral microcirculation during fluctuations in arterial pressure. In this chapter, structural and functional aspects of the cerebral circulation will be reviewed, including many of its unique properties. Given the large amount of subject matter, however, not all aspects of this unique circulation will be covered in detail.

CHAPTER 2

Anatomy and Ultrastructure

THE ARTERIES

The brain is one of the most highly perfused organs in the body. It is therefore not surprising that the arterial blood supply to the human brain consists of two pairs of large arteries, the right and left *internal carotid* and the right and left *vertebral arteries* (Figure 1). The internal carotid arteries principally supply the cerebrum, whereas the two vertebral arteries join distally to form the *basilar artery*. Branches of the vertebral and basilar arteries supply blood for the cerebellum and brain stem. Proximally, the basilar artery joins the two internal carotid arteries and other communicating arteries to form a complete anastomotic ring at the base of the brain known as the *circle of Willis*, named after

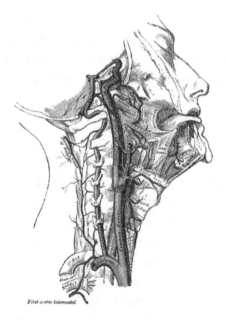

FIGURE 1: The internal carotid and vertebral arteries: right side. Reproduction of a lithograph plate from *Gray's Anatomy* from the 20th U.S. edition of *Gray's Anatomy of the Human Body*, originally published in 1918. It is not copyrightable in the United States as per Bridgeman Art Library v. Corel Corp.

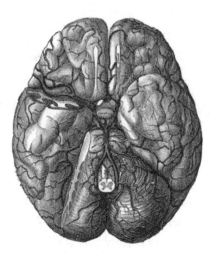

FIGURE 2: The arteries of the base of the brain. The temporal pole of the cerebrum and a portion of the cerebellar hemisphere have been removed on the right side. Reproduction of a lithograph plate from *Gray's Anatomy* from the 20th U.S. edition of *Gray's Anatomy of the Human Body*, originally published in 1918.

Sir Thomas Willis who described the arterial circle (*circulus arteriosus cerebri*). The circle of Willis gives rise to three pairs of main arteries, the *anterior*, *middle*, and *posterior cerebral arteries*, which divide into progressively smaller arteries and arterioles that run along the surface until they penetrate the brain tissue to supply blood to the corresponding regions of the cerebral cortex (Figure 2).

CEREBRAL VASCULAR ARCHITECTURE

The pial vessels are intracranial vessels on the surface of the brain within the pia–arachnoid (also known as the leptomeninges) or glia limitans (the outmost layer of the cortex comprised of astrocytic end-feet) [1]. Pial vessels are surrounded by cerebrospinal fluid (CSF) and give rise to smaller arteries that eventually penetrate into the brain tissue (Figure 3). Penetrating arterioles lie within the Virchow–Robin space and are structurally between pial and parenchymal arterioles. The Virchow–Robin space is a continuation of the subarachnoid space and varies considerably in depth by species [1]. The penetrating arteries become parenchymal arterioles once they penetrate into the brain tissue and become almost completely surrounded by astrocytic end-feet [2,3].

There are several important structural and functional differences between pial arteries on the surface of the brain and smaller parenchymal arterioles. First, pial arteries receive perivascular innervation from the peripheral nervous system also known as "extrinsic" innervation, whereas parenchymal arterioles are "intrinsically" innervated from within the brain neuropil (see *Perivascular*

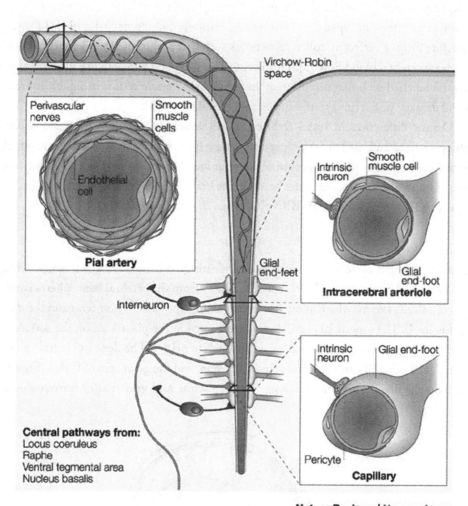

Nature Reviews | Neuroscience

FIGURE 3: Pial arteries on the brain surface have perivascular nerves that give rise to penetrating arteries within the Virchow–Robin space. As penetrating arterioles become parenchymal arterioles within the brain neuropil, they become associated with neurons and astrocytes. Parenchymal arterioles supply the cerebral microcirculation, known as the neurovascular unit. Used with permission from *Nat Rev Neurosci* 2004;5:347–360.

Innervation). While parenchymal arterioles have only one layer of circumferentially oriented smooth muscle, they possess greater basal tone and are unresponsive to at least some neurotransmitters that can have large effects on upstream vessels (e.g., serotonin, norepinephrine) [4]. Lastly, pial vessel architecture forms an effective collateral network such that occlusion of one vessel does not appreciably decrease cerebral blood flow [5]. However, penetrating and parenchymal arterioles are long and largely unbranched such that occlusion of an individual arteriole results in significant reductions in flow and damage (infarction) to the surrounding local tissue [5].

Despite differences in vessel architecture, all vessels in the brain have endothelium that is highly specialized and has barrier properties that are in some ways more similar to epithelium than endothelium in the periphery. Because of these unique barrier properties that tightly regulate exchange of nutrients, solutes, and water between the brain and the blood, the cerebral endothelium in known as the blood–brain barrier (BBB) [6,7] (see *Blood–Brain Barrier*).

THE VEINS

The cerebral venous system is a freely communicating and interconnected system comprised of dural sinuses and cerebral veins [8,9]. Venous outflow from the cerebral hemispheres consists of two groups of valveless veins, which allow for drainage: the *superficial cortical veins* and the *deep or central veins* (Figure 4). The superficial cortical veins are located in the pia matter on the surface of the cortex and drain the cerebral cortex and subcortical white matter. The deep or central veins consist of subependymal veins, internal cerebral veins, basal vein, and the great vein of Galen (Figure 5). These veins drain the brain's interior, including the deep white and gray matter surrounding the lateral

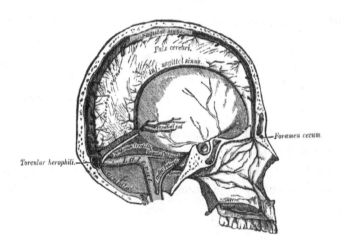

FIGURE 4: Superficial cortical veins and dural sinuses. Reproduction of a lithograph plate from *Gray's Anatomy* from the 20th U.S. edition of *Gray's Anatomy of the Human Body*, originally published in 1918.

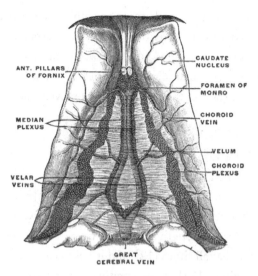

FIGURE 5: Deep or central veins. Reproduction of a lithograph plate from *Gray's Anatomy* from the 20th U.S. edition of *Gray's Anatomy of the Human Body*, originally published in 1918.

and third ventricles or the basal cistern and anastomose with the cortical veins, emptying into the *superior sagittal sinus* (SSS). Venous outflow from the SSS and deep veins is directed via a confluence of sinuses toward the sigmoid sinuses and jugular veins. The cerebellum is drained primarily by two sets of veins, the *inferior cerebellar veins* and the *occipital sinuses*. The brain stem is drained by the veins terminating in the inferior and transverse petrosal sinuses.

STRUCTURE OF CEREBRAL VESSELS

The wall of cerebral arteries and arterioles consist of three concentric layers: the innermost layer is the *tunic intima*, which consists of a single layer of endothelial cells and the internal elastic lamina (IEL); the next layer out is the *tunica media*, which contains mostly smooth muscle cells with some elastin and collagen fibers; and the outermost layer is the *tunica adventitia*, composed mostly of collagen fibers, fibroblasts, and associated cells such as perivascular nerves (in large and small pial arteries) and pericytes and astrocytic end-feet (in parenchymal arterioles and capillaries). Unlike systemic arteries, cerebral arteries have no external elastic lamina, but instead have a well-developed IEL [10]. Other differences from systemic arteries include a paucity of elastic fibers in the medial layer and a very thin adventitia. The number of smooth muscle cell layers varies depending on the size of the vessels and species, with large arteries such as the internal carotid artery having as many as 20 layers. Smaller pial arteries contain approximately two to three layers of smooth muscle, whereas the penetrating and parenchymal arterioles contain just one layer of smooth muscle. In addition, smooth muscle in the medial layer of cerebral arteries and arterioles are circularly arranged and oriented perpendicular to blood flow with essentially a zero-degree pitch. Cerebral veins are very thin-walled compared to arteries. The larger pial veins have circumferentially oriented smooth muscle that is not present in veins in the parenchyma. Unlike veins in the periphery, cerebral veins do not contain valves [9].

THE MICROCIRCULATION AND THE "NEUROVASCULAR UNIT"

The capillary bed of the brain is comprised of a dense network of intercommunicating vessels that consist of specialized endothelial cells and no smooth muscle [2]. The total length of capillaries in the human brain is ~400 miles [11]. It is the primary site of oxygen and nutrient exchange, which in turn is dependent on the path length and transit time of red blood cells. In the brain, all capillaries are perfused with blood at all times [12], and it has been estimated that nearly every neuron in the brain has its own capillary [13], demonstrating the critical relationship between the neuronal and vascular compartments. The intravascular pressure gradient between the precapillary arteriole and postcapillary venule is the primary regulator of capillary flow. Dilatation of resistance arteries and arterioles increases the microvascular pressure gradient and increases capillary flow. Thus, regulation of flow in the microcirculation is dependent on the regulation of flow and microvascular pressure in the brain arterioles. Red cell velocity in the cerebral capillary microcirculation is remarkably high (~1 mm/sec) and heterogeneous (range: 0.3 to 3.2 mm/sec) [14]. The heterogeneous flow velocity is important for effective oxygen transport to neuronal tissue that has considerable metabolic needs that fluctuate regularly.

Under normal conditions, the density of brain capillaries varies significantly within the brain depending on location and energy needs with higher capillary density in gray vs. white matter [15]. Pathological, physiological, and environmental states can influence or promote changes in capillary density. For example, chronic hypoxia increases capillary density through activation of angiogenic pathways (e.g., hypoxia inducible factor-1 and vascular endothelial growth factor) driven by a decrease in the driving force of Po_2 [16,17]. Brain capillary density nearly doubles between 1 and 3 weeks of chronic hypoxic exposure [16]. This adaptive increase in capillary density during chronic hypoxia increases cerebral blood volume [18] and restores tissue oxygen tension [19]. Hypertension also affects brain capillary density. Similar to the peripheral microcirculation, hypertension causes rarefaction (decrease in number) of capillaries and impaired microvessel formation that can increase vascular resistance [20].

Brain capillary structure is also unique compared to other organs. Endothelial cells and pericytes are encased by basal lamina (~30–40 nm thick) containing collagen type IV, heparin sulfate proteoglycans, laminin, fibronectin, and other extracellular matrix proteins [12,21]. The basal lamina of the brain endothelium is continuous with astrocytic end-feet that ensheath the cerebral capillaries (Figure 6). Astrocytes have a significant influence on capillary function, including regulating cerebral blood flow, upregulating tight junction proteins, contributing to ion and water homeostasis, and interfacing directly with neurons [2,3,12,22,23]. Although the barrier properties of the BBB are at the level of the tight junction in endothelial cells (see *Blood–Brain Barrier*), there is an important role for other components of the BBB, including the basement membrane, pericytes, astrocytes, and neurons. There is complex cross-talk between all entities and cell types, collectively known as the "neurovascular unit." Consideration of the neurovascular unit is important for disease processes that induce hemorrhage, vasogenic edema, infection, and inflammation [12,21,22,23]. The neurovascular unit may be the primary site of dysfunction for some disease state; however, for others such as atherosclerosis, large arteries are predominantly affected. For others, such as chronic hypertension, all segments of the circulation are affected.

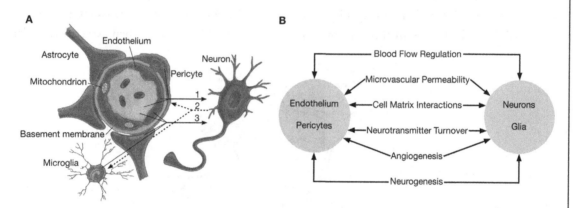

FIGURE 6: Schematic of the neurovascular unit. (A) Endothelial cells and pericytes are separated by the basement membrane. Pericyte processes sheathe most of the outer side of the basement membrane. At points of contact, pericytes communicate directly with endothelial cells through the synapse-like peg-socket contacts. Astrocytic end-foot processes unsheathe the microvessel wall, which is made up of ve a primary vascular origin , circulating neurotoxins may cross the BBB to reach their neuronal targets, or proinflammatory signals from the vascular cells or reduced capillary blood flow may disrupt normal synaptic transmission and trigger neuronal injury (arrow 1). Microglia recruited from the blood or within the brain and the vessel wall can sense signals from neurons (arrow 2). Activated endothelium, microglia, and astrocytes signal back to neurons, which in most cases aggravate the neuronal injury (arrow 3). In the case of a primary neuronal disorder, signals from neurons are sent to the vascular cells and microglia (arrow 2), which activate the vasculo-glial unit and c endothelial cells and pericytes. Resting microglia have a "ramified" shape. In cases of neuronal disorders that haontributes to the progression of the disease (arrow 3). (B) Coordinated regulation of normal neurovascular functions depends on the vascular cells (endothelium and pericytes), neurons, and astrocytes. Used with permission from *Neuron* 2008;57:178–201.

Pericytes

Pericytes were discovered in 1890 by Rouget as cells adjacent to capillaries that share a common basement membrane with endothelial cells [24]. The pericyte/endothelia ratio is high in the brain compared to the vasculature of other organs, e.g., 1:3 in brain vs. 1:100 in skeletal muscle [25]. Pericytes can be oriented along a blood vessel or circumvent the vessel with long processes that cover a large part of the abluminal surface. Pericytes have a number of potential roles in the brain, although it has been difficult to define these roles in vivo. They contribute to the stability of the vessel and release growth factors and matrix important for microvascular permeability, remodeling, and angiogenesis [26].

COLLATERALS

The collateral circulation in the brain consists of vascular networks that allow for maintenance of cerebral blood flow when principal inflow conduits fail due to occlusion or constriction. The circle of Willis at the base of the brain allows for redistribution of blood flow when extracranial or large

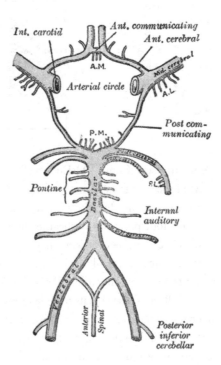

FIGURE 7: Diagram of the arterial circulation at the base of the brain. Reproduction of a lithograph plate from *Gray's Anatomy* from the 20th U.S. edition of *Gray's Anatomy of the Human Body*, originally published in 1918.

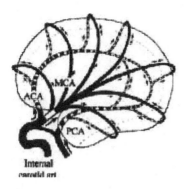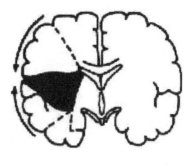

FIGURE 8: Collateral circulation of the brain. Heubner's leptomeningeal anastomoses connect the peripheral branches of the brain arteries and provide collateral blood flow to the peripheral parts of the adjacent vascular territories. Used with permission from *Cell Mol Neurobiol* 2006;26:1057–1083.

intracranial vessels are occluded [27,28] (Figure 7). This anastomotic loop provides low-resistance connections that allow reversal of blood flow to provide primary collateral support to the anterior and posterior circulations. However, the anatomy of the circle of Willis varies substantially with species and individuals and is often asymmetric [28].

The pial network of leptomeningeal vessels comprises secondary collaterals and are responsible for redistribution of flow when there is constriction or occlusion of an artery distal to the circle of Willis [27]. These vessels comprise distal anastomoses from branches of the anterior, middle, and posterior cerebral pial arteries (Figure 8). The functional capacity for collateral supply is dependent on the number and luminal caliber of the vessel that can be quite variable in the leptomeningeal anastomoses. Venous collaterals exist as well to augment drainage when primary routes are occluded or during venous hypertension [28]. The superficial cerebral veins are highly anastomosed with each other to provide a network of collaterals [8]. The deep veins are anastomosed with other venous systems and also provide collateral support for drainage [8].

CHAPTER 3

Perivascular Innervation

The large and small pial arteries and arterioles contain perivascular nerves within their adventitial layer that originate from the peripheral nervous system and are therefore considered "extrinsic" in nature. These perivascular nerves originate mainly from the superior cervical ganglia (SCG), the sphenopalatine (SPG), otic (OG), or trigeminal ganglion (TG) [22] (Figure 9). In pial vessels, extrinsic nerves form a network of varicose fibers within the adventitial layer (Figure 10) that decreases in density upon entering the Virchow–Robin space and then disappear in vessels within the brain parenchyma [22,29–31]. Parenchymal arterioles and cortical microvessels are innervated from within the brain tissue, and therefore, innervation is considered "intrinsic" in nature. These vessels receive nerve afferents from subcortical neurons from the locus coeruleus, raphe nucleus, basal forebrain, or local cortical interneurons that project to the perivascular space surrounding the arteriole [32–34]. This type of innervation, however, provides minimal contact with the parenchymal arterioles, but rather targets the surrounding astrocytes [2,12,22,32]. Careful ultrastructural studies found that very few (~7%) norepinephrine terminals from within the neuropil are actually associated with the vascular wall of parenchymal arterioles and were rarely junctional at the ultrastructural level [32]. The target of most terminals are astrocytes surrounding the arterioles in the brain neuropil [32,35,36]. Similarly, astrocytes express numerous subtypes of serotonin (5-HT) receptors that are involved in a variety of brain functions and diseases including depression and migraine [37–39]. Astrocytic function can also significantly affect vascular responses in the brain parenchyma (see *Astrocyte Control of Blood Flow*).

The differential innervation between pial vessels and parenchymal arterioles is reflected in a different population of post-synaptic receptors on the smooth muscle and endothelium. For example, the major post-junctional norepinephrine receptors in the rat middle cerebral artery smooth muscle are of the α1-adrenoceptor subtype [40–42] that cause vasoconstriction via activation of the phospholipase C-protein kinase C pathway [43]. This α-adrenoceptor reactivity becomes progressively less pronounced in the smaller pial arteries and is absent from the parenchymal arterioles due to a shift in receptor population from α- to β-adrenoceptor in the parenchymal arterioles that cause vasodilation [41,44–47]. Similar heterogeneity has been shown with 5-HT. The marked contractile

FIGURE 9: Cerebral artery stained with the pan-neuronal stain protein gene product 9.5. Varicose nerve fibers are evident along the length of the vessel. Magnification 40×.

effect of the large cerebral pial arteries, due to stimulation of post-junctional 5-HT1B receptors, is absent from small arterioles that dilate instead of constrict to the neurotransmitter [41,48–50].

The existence of perivascular innervation of the cerebral circulation has been known for centuries; however, the functional significance of this innervation is still poorly understood [51]. Extrinsic innervation of the pial vasculature has considerable heterogeneity in both the pattern and density of innervation depending on the location within the brain. For example, nerve density for both 5-HT and norepinephrine is increased rostral vs. caudal, such that the internal carotid artery system is more densely innervated than the vetebrobasilar [52,53]. This differential perivascular nerve density may serve to affect the degree of vasoconstriction and the differential metabolic needs of each particular vascular territory in the brain [52,54]. To that end, there is functional heterogeneity in the response of cerebral arteries to neurotransmitter. For example, reactivity to norepinephrine is greater in anterior compared to middle cerebral arteries of cats, suggesting differential post-synaptic receptor density and/or composition between arteries supplying different brain regions (middle vs. anterior) [55].

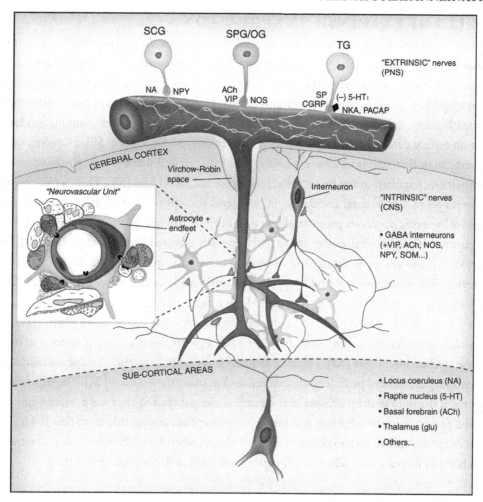

FIGURE 10: Schematic representation of the different types of perivascular nerves. The "extrinsic" nerves to cerebral blood vessels at the surface of the brain come from the peripheral nervous system (PNS) and originate in the superior cervical (SCG), sphenopalatine (SPG), or otic (OG) or trigeminal (TG) ganglion. Blood vessels located within the brain parenchyma, or the microcirculation, are innervated by "intrinsic" nerve pathways that find their origin in the central nervous system (CNS). For cortical microvessels, anatomical and/or functional evidence indicate that they receive NA, 5-HT, ACh, or GABAergic afferents from subcortical neurons from the locus coeruleus, raphe nucleus, basal forebrain, or local cortical interneurons. ACh, acetylcholine; CGRP, calcitonin gene-related peptide; GABA, γ-aminobutyric acid; NA, norepinephrine; NKA, neurokinin-A; NOS, nitric oxide synthase; NPY, neuropeptide Y; PACAP, pituitary adenylate-cyclase activating polypeptide; SOM, somatostatin; SP, substance P; VIP, vasoactive intestinal polypeptide; 5-HT, serotonin. Used with permission from *J Appl Physiol* 2006;100:1059–1064.

ROLES OF EXTRINSIC INNERVATION

Cerebral pial vessels are innervated by perivascular nerves that originate peripherally from both autonomic and sensory ganglia, including sympathetic, parasympathetic, and trigeminal. Innervation of cerebral arteries by the sympathetic nervous system arises from the SCG with norepinephrine and neuropeptide Y as primary neurotransmitters [22]. Despite the significant vasoconstrictor effects of sympathetic neurotransmitters in large cerebral arteries [41], sympathetic stimulation has little effect on resting cerebral blood flow under normotensive conditions [56–58]. However, during acute hypertension that results in autoregulatory breakthrough and decreased cerebrovascular resistance, sympathetic stimulation significantly decreases cerebral blood flow [59]. Thus, a role of sympathetic innervation of the cerebral circulation is to extend the upper limit of autoregulation and protect against increases in venous pressure, disruption of the BBB, and edema formation that occurs during hypertensive encephalopathy due to pathologically elevated hydrostatic pressure [41,56,58]. Innervation from the parasympathetic nervous system arises mainly from the OG and SPG with acetylcholine, vasoactive intestinal peptide, and nitric oxide as neurotransmitters [22]. Stimulation of parasympathetic nerves has potent vasodilator effects on cerebral arteries, and increases cerebral blood flow [60,61]. Nitric oxide shifts the autoregulatory curve to the lower range of pressures [62] (see *Autoregulation of Cerebral Blood Flow*) and has a protective role during cerebral ischemia to increase cerebral blood flow [63]. Trigeminal innervation arises from the TG and contains substance P, calcitonin gene-related peptide, and neurokinin-A as neurotransmitters [22]. The trigeminal nervous system is the only sensory afferent in the cerebral circulation and thus has been strongly implicated in the pathogenesis of migraine and mediates nociception during this condition [64,65]. Stimulation of the trigeminal system does not affect resting cerebral blood flow, but has protective effects in the brain to increase flow when it is compromised such as during vasospasm [66].

CHAPTER 4

Regulation of Cerebrovascular Tone

MYOGENIC RESPONSE

Discovered over 100 years ago by Bayliss, the myogenic response is the intrinsic property of smooth muscle to respond to changes in mechanical load or intravascular pressure [67]. It is a critical component of resistance artery function and is more prominent in the cerebral circulation than many other vascular beds. The smooth muscle of both large arteries and small arterioles constrict in response to increased pressure and dilate in response to decreased pressure (known as the "Bayliss effect"), thus contributing to autoregulation of blood flow [68,69]. This innate myogenic activity is also crucial for normal hemodynamic function and for maintaining vascular resistance, which serves to protect smaller downstream arterioles and capillaries from damage in the face of changing perfusion pressures and to maintain tissue perfusion during periods of decreased blood pressure [68,69]. The myogenic response arises from smooth muscle and exists in arteries and arterioles denuded of endothelium and in sympathetically denervated animals [70]. Thus, the myogenic response is truly myogenic in nature. However, release of vasoactive factors from both endothelium and perivascular nerves and local metabolites can increase or decrease the level of myogenic tone and thus affect vascular resistance.

The myogenic behavior of resistance arteries and arterioles involves two phenomenon: *myogenic tone*, which is a state of partial constriction at a constant pressure, and *myogenic reactivity*, which is the alteration in tone in response to a change in pressure [71] (Figure 11). A third phase also occurs at excessively high arterial pressure beyond the autoregulatory pressure range and involves marked increases in vessel diameter and loss of tone that occurs during autoregulatory breakthrough, known as *forced dilatation* [72]. Although the term "forced dilatation" implies this phenomenon is a mechanical event, it likely is an active vasodilation possibly involving K_{Ca} channels and/or reactivity oxygen species, invoked to protect the arterial wall from damage [73]. Mechanisms underlying both the development of myogenic tone and the more physiological myogenic reactivity are not completely understood and appear to vary between species and vascular beds. However, for the sake of simplicity, it is worth considering the phases of the response as having different triggers and mechanisms of regulation, although it is likely that they are highly interactive and interdependent.

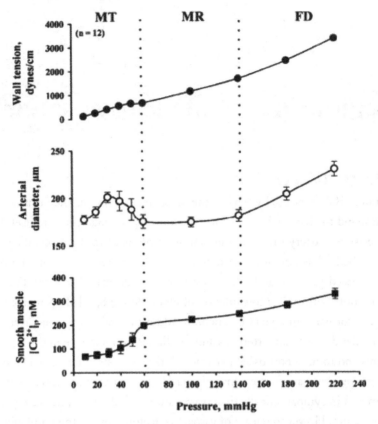

FIGURE 11: Graph summarizing changes in wall tension, arterial diameter, and arterial wall $[Ca^{2+}]_i$ in response to stepwise increase in transmural pressure from 10 to 220 mmHg. Vertical dotted lines partition the pressure range in accordance with the three phases of the model (MT, MR, and FD). Used with permission from *Am J Physiol* 2002;283:H2260–H2267.

Mechanisms of Myogenic Response Initiation

The initiation or development of the myogenic response occurs via ionic and enzymatic mechanisms, both of which increase intracellular calcium [74]. Increased pressure causes depolarization of the smooth muscle cell membrane and calcium influx via opening of voltage-operated calcium channels (Ca_v) [75,76]. In particular $Ca_v1.2$ are the prominent calcium channel involved, but in some segments of the vasculature such as small parenchymal arterioles, other Ca_v channels may participate [76]. The rise in intracellular calcium increases myosin light-chain (MLC) phosphorylation and promotes vasoconstriction. Removal of extracellular calcium abolishes the myogenic response, suggesting a prominent role for calcium influx in myogenic response initiation [74,75]. Although a role for Ca_v channels in the initiation of the myogenic response is well-established, the primary stimulus or sensor that transduces the change in pressure (a mechanical event) into depolarization and vasoconstriction are not clear. Wall tension has been proposed as a stimulus for initiation of the myogenic response because it has been shown to correlate with changes in cell calcium and MLC phosphorylation, a relationship not seen with diameter [77,78]. Stretch-activated cation channels, including transient receptor potential (TRP) channels, are thought to be the sensor by which pressure causes depolarization [79,80], although Ca_v channels may also be activated directly by transmural pressure [81]. In addition, chloride channels may also participate in pressure-induced smooth muscle depolarization because their activation increases an inward current, causing depolarization [82]. In addition to ionic mechanisms, there is evidence for other factors involved in mechanotransduction during the development of myogenic tone including integrins and actin cytoskeletal dynamics [83,84]. Both integrins and TRP channels are linked to the actin cytoskeleton [85,86], providing a connection by which all these process can interact to transduce pressure or stretch into a depolarization and contractile response.

Mechanisms of Myogenic Reactivity

The vasoconstriction associated with the development of tone is different in many aspects than the response of the vessel to pressure once tone is present. During this phase, an increase in intravascular pressure does not change vessel diameter significantly and can cause further constriction [71,84]. The vessel wall stiffens due to enhanced MLC phosphorylation and contraction that is further reinforced by actin polymerization [77,84,87]. It is unlikely that tension or stretch is a stimulus for constriction during this phase because wall tension is actually increased due to elevated pressure, not decreased as the model predicts with tone development. Another fundamental difference from tone initiation is that there is little change in smooth muscle calcium, but calcium sensitivity is enhanced during myogenic reactivity [71,73,88]. When calcium is clamped by exposure to high extracellular calcium or in permeabilized artery preparations, myogenic reactivity is still present [71,89]. Several mechanisms exists by which pressure induces calcium sensitivity in smooth muscle, including activation of protein kinase C (PKC) and RhoA-Rho kinase pathways [71,73,89,90] (Figure 12).

Feedback

At least one major negative feedback mechanism exists that limits the myogenic vasoconstriction. Calcium-activated potassium channels (K_{ca}), in particular large-conductance K_{Ca} or BK_{Ca} channels, expressed on cerebral artery smooth muscle, are activated by intracellular calcium release

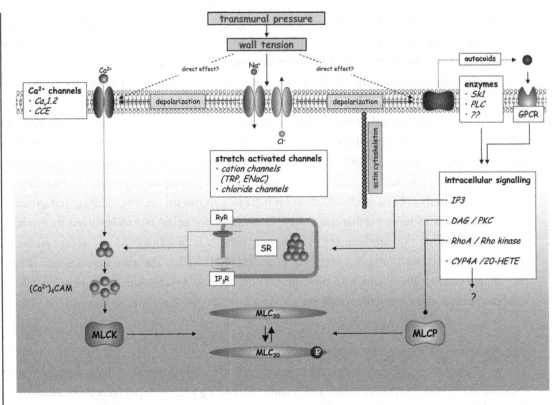

FIGURE 12: Signaling pathways in vascular smooth muscle reported to contribute to the myogenic response. Schematic representation depicting the primary signaling pathways that are currently believed to initiate and maintain the myogenic response. Of significance, the represented pathways (e.g., "Ca^{2+}-dependent" (left) and "Ca^{2+}-independent" (right)) are not mutually exclusive. In fact, there is growing evidence that the myogenic response possesses a high degree of redundancy at the mechanosensor and intracellular signaling pathway levels. For details, please refer to the main text. Abbreviations: TRP, transient receptor protein cation channel; ENaC, degenerin/epithelial sodium cation channel; Cav1.2, voltage-operated Ca^{2+} channel; CCE, capacitative calcium entry; $(Ca^{2+})4CAM$, Ca^{2+}-calmodulin complex; MLCK, myosin light-chain kinase; MLC20, myosin light-chain regulatory domain (20 kDa); Sk1, sphingosine kinase 1; PLC, phospholipase C; GPCR, G-protein-coupled receptor; IP3, inositol triphosphate; DAG, diacylglycerol; PKC, protein kinase C; IP3R, IP3 receptor; RyR, ryanodine receptor; sER, smooth endoplasmic reticulum; CYP4A, cytochrome P450 4A; MLCP, myosin light-chain phosphatase. Used with permission from *Cardiovasc Res* 2008;77:8–18.

events or "calcium sparks" whose frequency is regulated by transmural pressure [91–93]. Activation of BK_{Ca} channels by calcium sparks causes hyperpolarization and attenuation of the myogenic vasoconstriction [91,92]. This negative feedback mechanism serves to put the brakes on the constriction induced by pressure-induced depolarization and increased intracellular calcium [73,93].

Effect of Disease States on Myogenic Tone and Reactivity

The myogenic response has a prominent role in normal hemodynamic processes in the brain. The basal constriction due to myogenic mechanisms provides a state from which an artery or arteriole can increase or decrease diameter on demand, thereby modulating cerebrovascular resistance and contributing to local and global blood flow regulation [68,69,71,72]. Conducted or flow-mediated vasodilation of upstream vessels associated with functional hyperemia may involve myogenic vasodilation in response to decreased intravascular pressure [94]. The importance of the myogenic response in the brain is demonstrated by numerous disease states in which myogenic mechanisms are dysregulated, causing secondary brain injury such as ischemia and vasogenic edema [95,96]. For example, during focal ischemia when a thrombus or emboli occludes a cerebral vessel, there is both a decrease in flow and pressure that both contribute to autoregulation. Decreased flow causes hypoxia that when severe can promote vasodilation by metabolic mechanisms [68], whereas decreased pressure causes myogenic vasodilation. The decreased tone, due to both metabolic and myogenic involvement, diminishes cerebrovascular resistance, which can cause vasogenic edema formation due to significantly elevated hydrostatic pressure on the microcirculation [97] (see *Vasogenic Edema Formation*).

ENDOTHELIAL REGULATION OF TONE

The endothelium is a highly specialized cell type in the brain [12,23]. Similar to peripheral organs, it is involved in numerous physiological processes, including regulation of inflammatory and immune responses, thrombosis, adhesion, angiogenesis, and permeability [12] (see *Barriers of the CNS*). The importance of the endothelium is demonstrated by the fact that endothelial dysfunction has a central role in the pathogenesis of several cerebrovascular diseases such as Alzheimer's disease, epilepsy, and stroke [12,23,98–100]. The endothelium can produce several vasoactive mediators, including nitric oxide (NO), prostacyclin, and endothelium-derived hyperpolarizing factor (EDHF) that have a significant influence on vascular tone and thereby influence cerebral blood flow [101] (Figure 13).

Nitric Oxide

NO has some of the most prominent effects in the brain and is one of the most studied vasodilators. Under resting conditions, basal NO production by the cerebral endothelium inhibits resting tone in both large and small pial arteries and parenchymal arterioles and thus affects CBF [102]. NO

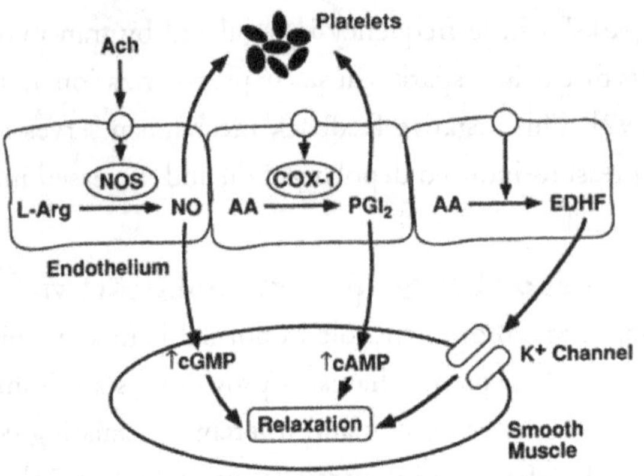

FIGURE 13: Some mechanisms of endothelium-dependent relaxation of cerebral vascular muscle. Nitric oxide (NO) is produced by NO synthase (NOS) from amino acid L-arginine (L-Arg). NO diffuses to vascular muscle where it activates soluble guanylate cyclase, causing increased production of guanosine 3′,5′-cyclic monophosphate (cGMP), which results in relaxation. Prostacyclin (PGI2) is normally produced by cyclooxygenase-1 (COX-1) from arachidonic acid (AA). PGI2 diffuses to vascular muscle where it activates adenylate cyclase, causing increased production of adenosine 3′,5′-cyclic monophosphate (cAMP), which results in relaxation. Endothelium-derived hyperpolarizing factor (EDHF) is probably a product of AA metabolism. EDHF diffuses to vascular muscle where it activates potassium (K+) channels. Increased activity of potassium channels produces hyperpolarization and relaxation of vascular muscle. ACh, acetylcholine. Used with permission from *Physiol Rev* 1998;78:53–97.

synthase (NOS) is the enzyme responsible for the oxygen-dependent conversion of L–arginine to NO and L-citrulline [103]. Three isoforms of NOS exist in brain with neuronal (nNOS) and endothelial (eNOS) constitutively expressed in neurons and cerebral endothelium, respectively [104]. Inducible (iNOS) is not normally expressed in brain, but its expression can be induced under certain pathological conditions [104]. The significant influence of NO on basal tone is demonstrated by effects of compounds that inhibit NOS, such as L-nitro-arginine (L-NNA), that cause significant endothelium-dependent vasoconstriction of cerebral arteries and arterioles [101,102,105].

NO production by the cerebral endothelium can be induced by stimuli that increase endothelial cell intracellular calcium (e.g., acetylcholine, ACH, shear stress) since NOS is calcium-dependent [106]. Production of NO by eNOS is principally activated by calcium-dependent binding of calmodulin, making eNOS activation calcium-dependent. Thus, many factors that increase intracellular calcium activate eNOS. Whether NO is produced basally or is induced, NO diffuses to VSM where it causes vasodilation primarily by activation of soluble guanylyl cyclase (sGC) [106].

Activation of sGC increases the levels of cyclic guanine monophosphate (cGMP) that in turn acti-vates protein kinase G, causing relaxation of VSM in part by opening BK_{Ca} channels and reducing intracellular calcium [107].

A critical determinant of eNOS activity is the availability of the cofactor tetrahydrobiopterin (BH_4) (108). Under conditions of limited BH_4 availability (due to oxidation or reduced formation), eNOS can function in an uncoupled state so that NAD(P)H-derived electrons are added to molecu-lar oxygen instead of L-arginine, leading to the production of superoxide [109]. Uncoupling of eNOS has been implicated in a number of diseases associated with decreased BH_4 levels, including athero-sclerosis, diabetes, and hypertension [110–113]. eNOS activity is also regulated by phosphorylation of the enzyme on specific amino acid residues [114,115]. Following Ser1177 phosphorylation, NO production is increased. In contrast, eNOS activity is inhibited by Thr495 phosphorylation [115].

Endothelium-Derived Hyperpolarizing Factor

Experimentally, EDHF is defined as the residual vasodilator mechanism that remains in the pres-ence of inhibition of NOS and cyclooxygenases (COX). EDHF-mediated vasodilation potentially occurs under basal conditions, but also in response to agonist-induced increase in endothelial cell calcium and activation of small- and intermediate-conductance calcium-activated potassium chan-nels or SK_{Ca} and IK_{Ca} channels (also known as $K_{Ca2.3}$ and $K_{Ca3.1}$, respectively) that hyperpolar-izes endothelial cells [116,117]. Endothelial cell hyperpolarization is then transferred to adjacent smooth muscle causing closure of Ca_v channels and relaxation (Figure 14). How hyperpolarization is transferred from endothelium to smooth muscle is not clear, but may involve gap junctions, which communicate through direct electrical coupling or by passage of chemical mediators [118,119]. Proposed chemical mediators include K^+ ions, cytochrome P450 metabolites, or epoxyeicosatri-enoic acid (EETs) [120–122].

The hyperpolarizing influence of EDHF appears to depend on the activity of SK_{Ca} and IK_{Ca} channels that are expressed in cerebral endothelium, but not smooth muscle [105,116,117]. Com-bined and specific blockade of these channels abrogates EDHF responses [105,116,117]. However, while activation of SK_{Ca} and IK_{Ca} channels by endothelial cell calcium is a defining feature of EDHF production, the precise contribution of each channel in cerebral vessels to the response is less clear. Agonist-induced increase in endothelial cell calcium increases the open probability of both channels, but inhibitor sensitivity and genetic deletion of IK_{Ca} channels indicate that the opening of IK_{Ca} chan-nels plays a greater role in mediating hyperpolarization and EDHF-dependent relaxation [123,124]. Although inhibition of SK_{Ca} and IK_{Ca} channels abolishes stimulated EDHF (i.e., agonist-induced in the presence of NOS and COX inhibition), it does not have direct effects on tone of larger cerebral arteries. In contrast, parenchymal arterioles constrict to both SK_{Ca} and IK_{Ca} channel inhibition [105],

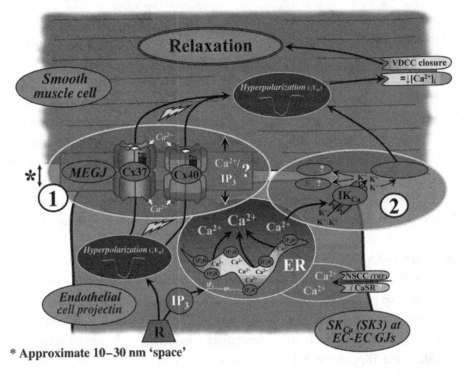

FIGURE 14: Mechanism of endothelium-derived hyperpolarization (EDH): endothelial to smooth muscle signaling in rat mesenteric artery. In response to agonist (R), intracellular endothelial cell (EC) calcium release occurs from inositol 1,4,5-trisphosphate (IP3) receptor (IP3R)-mediated [29,30] endoplasmic reticulum (ER) stores, which are in close proximity to myoendothelial gap junction (MEGJ) connexin (Cx) [37] and Cx40 (mechanism 1) [32] and intermediate conductance calcium-activated potassium channels (IKCa; mechanism 2) [32]. The net result of such activity is hyperpolarization of the adjacent smooth muscle, closure of voltage-dependent calcium channels (VDCC), subsequent reduced smooth muscle cell (SMC) calcium and vessel relaxation. Used with permission from *Clin Exp Pharmacol Physiol* 2009;36:67–76.

suggesting that in these small arterioles EDHF influences tone under basal conditions and thus influences resting cerebral blood flow.

The transfer of hyperpolarization from endothelium is thought to occur at structures called *myoendothelial junctions* (MEJ) [125]. MEJ are holes in the internal elastic lamina where there is close (10–30 nm) contact between endothelium and smooth muscle cell membranes [125] (Figure 15). Endothelium and smooth muscle cells communicate at these projections that act as pathways for diffusion of vasoactive substances between endothelium and smooth muscle or vice versa. Some communication at MEJ occurs through gap junctions known as myoendothelial gap junctions (MEGJ);

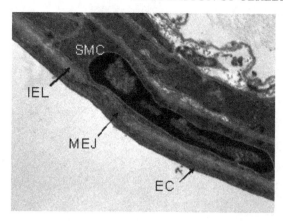

FIGURE 15: Transmission electron micrograph showing the presence of a MEJ in a parenchymal arteriole. Magnification 8000×.

however, not all MEJ contain gap junctions and the prevalence of MEJ increases with decreasing size of vessels [126,127].

Prostacyclin

In endothelial cells, arachidonic acid metabolism produces vasoactive products that are vasoconstricting and vasodilating in nature, although the involvement of this pathway in cerebral artery vasoactivity appears to be considerably less than in peripheral endothelium [127,128]. Activation of the calcium-dependent enzyme phospholipase A_2 hydrolyzes cellular lipid membranes to produce arachidonic acid lipid precursors that are substrates for cyclooxygenase (COX), lipoxygenases, and cytochrome P450 monooxygenases [128]. Products of the COX pathway can be vasodilating (prostaglandin I_2 (PGI_2), prostaglandin E_2 (PGE_2), prostaglandin D_2 (PGD_2)) or vasoconstricting (prostaglandin $F_{2\alpha}$ ($PGF_{2\alpha}$) and thromboxane A_2 (TXA_2)) in nature [127]. PGI_2 is synthesized from PGH_2 by PGI_2 synthase, which can activate adenylate cyclase and increase cyclic AMP and protein kinase A in smooth muscle, causing vasodilation [129]. Some prostaglandins are elevated in response to injury. TXA_2 causes vasoconstriction in response to hemorrhage [130,131]. Products of the lipoxygenase pathway are thought to be EDHFs in peripheral vessels, but not cerebral [132].

CHAPTER 5

Control of Cerebral Blood Flow

The brain uses ~20% of available oxygen for normal function, making tight regulation of blood flow and oxygen delivery critical for survival [133]. In a normal physiological state, total blood flow to the brain is remarkably constant due in part to the prominent contribution of large arteries to vascular resistance [58] (see *Segmental Vascular Resistance*). In addition, parenchymal arterioles have considerable basal tone and also contribute significantly to vascular resistance in the brain [58,105]. The high metabolic demand of neuronal tissue requires tight coordination between neuronal activity and blood flow within the brain parenchyma, known as functional hyperemia [21,22,134] (see *Neural–Astrocyte Regulation*). However, in order for flow to increase to areas within the brain that demand it, upstream vessels must dilate in order to avoid reductions in downstream microvascular pressure [58,135]. Therefore, coordinated flow responses occur in the brain, likely due to conducted or flow-mediated vasodilation from distal to proximal arterial segments and to myogenic mechanisms that increase flow in response to decreased pressure [94] (see *Myogenic Response*).

CEREBRAL HEMODYNAMICS

Brain blood flow can be modeled from a physical standpoint as flow in a tube with the assumptions that flow is steady, laminar, and uniform through thinned-walled (the wall is <10% of the lumen) non-distensible tubes [87]. These assumptions do not apply to large arteries that have thick walls or in the microcirculation in which flow is non-Newtonian [161]. Ohm's law states that flow is proportional to the difference in inflow and outflow pressure (ΔP) divided by the resistance to flow (R): flow=$\Delta P/R$. In the brain, ΔP is cerebral perfusion pressure (CPP), the difference between intra-arterial pressure and the pressure in veins. Venous pressure is normally low (2–5 mmHg) and is influenced directly by intracranial pressure (ICP). Therefore, ΔP is calculated as the difference in CPP and either venous pressure or ICP, whichever is greater. Blood flow is also estimated by Poiseuielle's law that states that flow is directly related to ΔP, blood viscosity, and the length of the vessel (assumed to be constant) and inversely related to radius to the fourth power: flow = $(8 \times \eta \times L)/r^4$ [136]. Thus, radius is the most powerful determinant of blood flow and even small changes in lumen diameter have significant effects on cerebral blood flow, and it is by this mechanism that vascular resistance can change rapidly to alter regional and global cerebral blood flow [137].

AUTOREGULATION OF CEREBRAL BLOOD FLOW

Autoregulation of cerebral blood flow is the ability of the brain to maintain relatively constant blood flow despite changes in perfusion pressure [137]. Autoregulation is present in many vascular beds, but is particularly well-developed in the brain, likely due to the need for a constant blood supply and water homeostasis. In normotensive adults, cerebral blood flow is maintained at ~50 mL per 100 g of brain tissue per minute, provided CPP is in the range of ~60 to 160 mmHg [138]. Above and below this limit, autoregulation is lost and cerebral blood flow becomes dependent on mean arterial pressure in a linear fashion [71,72,139]. When CPP falls below the lower limit of autoregulation, cerebral ischemia ensues [27,140]. The reduction in cerebral blood flow is compensated for by an increase in oxygen extraction from the blood [141]. Clinical signs or symptoms of ischemia are not seen until the decrease in perfusion exceeds the ability of increased oxygen extraction to meet metabolic needs. At this point, clinical signs of hypoperfusion occur, including dizziness, altered mental status, and eventually irreversible tissue damage (infarction) [140,141].

The mechanisms of autoregulation in the brain are not completely understood and likely differ with increases vs. decreases in pressure. Although a role for neuronal involvement in autoregulation is appealing, studies have shown that cerebral blood flow autoregulation is preserved in sympathetically and parasympathetically denervated animals, indicating that a major contribution of extrinsic neurogenic factors to autoregulation of cerebral blood flow is unlikely [70] (see *Perivascular Innervation*). Recently, a role for neuronal nitric oxide in modulating cerebral blood flow autoregulation has been shown, suggesting that although extrinsic innervation may not be involved, intrinsic innervation may have a role [62]. Biproducts of metabolism have also been proposed to have a role in autoregulation [142]. Reductions in cerebral blood flow stimulate release of vasoactive substances from the brain that cause arterial dilatation. Candidates for these vasoactive substances include H^+, K^+, O_2, adenosine, and others. Autoregulation of cerebral blood flow when pressure fluctuates at the high end of the autoregulatory curve is most likely due to the myogenic behavior of the cerebral smooth muscle that constrict in response to elevated pressure and dilate in response to decreased pressure [68,69–71]. The important contribution of myogenic activity to autoregulation is demonstrated in vitro in isolated and pressurized cerebral arteries that constrict in a response to increased pressure and dilate in response to decreased pressure [71,105] (see *Myogenic Response*). Autoregulation at pressures below the myogenic pressure range likely involves hypoxia and release of metabolic factors [68].

The importance of autoregulation in normal brain function is highlighted by the fact that significant brain injury occurs when autoregulatory mechanisms are lost. For example, during acute hypertension at pressures above the autoregulatory limit, the myogenic constriction of vascular smooth muscle is overcome by the excessive intravascular pressure and forced dilatation of cerebral vessels occurs [143–146]. The loss of myogenic tone during forced dilatation decreases cerebrovascular resistance, a result that can produce a large increase in cerebral blood flow (300–400%), known as autoregulatory breakthrough [143–146] (Figure 16). In addition, decreased cerebrovascular resistance increases hydrostatic pressure on the cerebral endothelium, causing edema formation [143–145], the underlying cause of conditions such as hypertensive encephalopathy, posterior reversible encephalopathy syndrome (PRES), and eclampsia [143,147] (see *Vasogenic Edema Formation*).

Although uncommon since the advent of effective antihypertensive therapy, hypertensive encephalopathy occurs as a result of a sudden, sustained rise in blood pressure sufficient to exceed the upper limit of cerebral blood flow autoregulation (>160 mmHg) [148–150]. Early studies on the reaction of cerebral vessels to high blood pressure produced the concept of hypertensive vasospasm. Acute hypertensive encephalopathy was thought the result of spasm—defined as an uncontrolled vasoconstriction—of the cerebral arteries, causing brain tissue ischemia [151,152]. This concept originated from the observations of Byrom [151] who produced experimental renal hypertension and found ~90% of hypertensive rats with neurologic manifestations showed multiple cortical spots of trypan blue extravasation, whereas rats without cerebral symptoms appeared to have normal cerebrovascular permeability. He also noted what he called an alternating vasoconstriction/vasodilation in the pial vessels, a phenomenon known as a "sausage-string" appearance. This observation led him

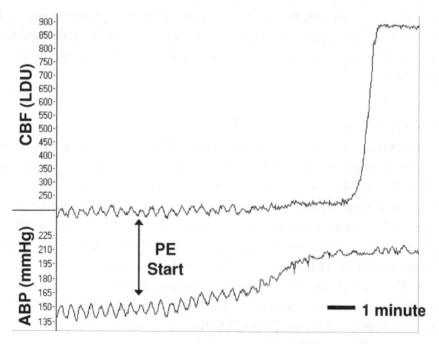

FIGURE 16: Tracing of CBF (in laser Doppler units) and ABP (in mmHg) in response to increasing doses of PE. In this experiment, CBF increased four times greater than baseline as ABP increased from 140 to 210 mmHg, demonstrating autoregulatory breakthrough. Used with permission from *Hypertension* 2007;49:334–340.

to the conclusion that cerebral vasospasm caused ischemia and edema formation in response to acute hypertension. Byrom later modified his view and referred to a finding in the mesenteric circulation that vessels with this "sausage-string" appearance had protein leakage in the dilated parts of the vessels only [153,154]. Since then, it has been established that high blood pressure results in increased cerebral blood flow and "breakthrough of autoregulation" [155]. Further experiments confirmed that loss of myogenic vasoconstriction during forced dilatation rather than spasm is the critical event in hypertensive encephalopathy [156].

SEGMENTAL VASCULAR RESISTANCE

In peripheral circulations, small arterioles (<100 μm diameter) are typically the major site of vascular resistance (157). However, in the brain, both large arteries and small arterioles contribute significantly to vascular resistance. Direct measure of the pressure gradient across different segments of the cerebral circulation found that the large extracranial vessels (internal carotid and vertebral) and intracranial pial vessels contribute ~50% of cerebral vascular resistance [58,158]. Large artery resistance in the brain is likely important to provide constant blood flow under conditions that change blood flow locally, e.g., metabolism. Large artery resistance also attenuates changes in downstream microvascular pressure during increases in systemic arterial pressure. Thus, *segmental vascular resistance* in the brain is a protective mechanism that helps provide constant blood flow in an organ with high metabolic demand without pathologically increasing hydrostatic pressure that can cause vasogenic edema.

NEURAL–ASTROCYTE REGULATION

Unlike pial arteries and arterioles, parenchymal arterioles are in close association with astrocytes and, to a lesser extent, neurons. Both these cell types may have a role in controlling local blood flow [2,12,22,32]. Subcortical microvessels are innervated from within the brain parenchyma and are unique in that the majority of vericosities adjoin astrocytic end-feet surrounding arterioles and thus does not have conventional neurovascular junctions [135]. Neurons whose cell bodies are from within the subcortical brain regions (e.g., nucleus basalis, locus ceruleus, raphe nucleus) project to cortical microvessels to control local blood flow by release of neurotransmitter (e.g., ACH, norepinephrine, 5HT) [22] (Figure 17). Release of neurotransmitter stimulates receptors on smooth muscle, endothelium, or astrocytes to cause constriction or dilation, thereby regulating local blood flow in concert with neuronal demand [22,98,134]. It has been known for some time that astrocytes can release vasoactive factors [159]. Evidence for the involvement of astrocytes in local control of blood flow in vivo has recently emerged. Their close apposition to microvessels, encasing almost the entire

SUB-CORTICAL AREAS	CEREBRAL CORTEX	VASOACTIVE MEDIATOR	RECEPTOR	VASOMOTOR RESPONSE
		NO, ACh, VIP,	–, M5, VPAC1	Dilatation
		GABA	GABA$_A$	Dilatation
		NPY, SOM	Y1, SSR2/4?	Contraction
		NO	–	Dilatation
		ACh	M5	Dilatation
		5-HT	5-HT$_{1B}$	Contraction
		PGE$_2$	EP4	Dilatation
		20-HETE	?	Contraction

FIGURE 17: Summary of the regulation of cortical microvessels from cells located in subcortical areas and within the cerebral cortex. The possibility that interneurons also induce the release of vasoactive molecules from astrocytes is not included for clarity purposes. The known or suggested vasoactive mediators and the vascular receptors on which neuronal or astroglial (PGE$_2$ and 20-HETE) signaling molecules are believed to act to induce dilatation or constriction are illustrated. Note that GABA has been shown to dilate, via GABAA receptors, pial vessels but not intracortical microvessels [12]. M5, muscarinic receptors that mediate dilatation of cerebral microvessels; VAPC1, dilatory receptor for VIP in brain vessels; NPY1, NPY receptor mediating cerebral vasoconstriction; SSR2/4, somatostatin receptors on smooth muscle cells of cortical microvessels that can mediate contraction [4]; 5-HT1B, contractile receptor for 5-HT, but note that a dilatory response mediated by the same receptor has also been reported [11]; EP4, dilatory receptors for PGE$_2$ in brain vessels [7]; ?, the cerebrovascular receptor for 20-HETE is still unknown. Used with permission from *J Appl Physiol* 2006;100:1059–1064.

parenchymal arterioles and capillaries with little neuronal contact, makes astrocytic involvement likely at this level [21,22,98,134]. Studies in brain slices, in which the entire neurovascular unit is intact, showed that direct electrical stimulation of neuronal processes raises calcium in astrocytic end-feet and causes dilation of nearby arterioles [160]. Stimulation of astrocytes also raises calcium in end-feet and has a similar vasoactive effect on parenchymal arterioles; however, whether dilation or constriction occurs seems to depend on the level of calcium and, not surprisingly, resting tone [161]. It has been proposed that an elevation in astrocyte calcium releases vasoactive factors, including K$^+$, 20-HETE, and PGE$_2$ [160–162]. However, a weakness of the brain slice preparation is that it does not allow for arterioles to be pressurized or have flow. Thus, the role of the myogenic response, which may significantly modify any astrocytic-derived signals in vivo is not known.

EFFECT OF OXYGEN

The brain has a very high metabolic demand for oxygen compared to other organs, and thus, it is not surprising that acute hypoxia is a potent dilator in the cerebral circulation that produces marked increases in cerebral blood flow [163]. In general, blood flow does not change in the brain until tissue Po_2 falls below ~50 mmHg, below which cerebral blood flow increases substantially [163]. As hypoxia decreases Po_2 further, cerebral blood flow can rise up to 400% of resting levels [164]. Increases in cerebral blood flow do not change metabolism, but hemoglobin saturation falls from ~100% at Po_2>70 mmHg to ~50% at Po_2<50 mmHg [164]. Acute hypoxia causes an increase in cerebral blood flow via direct effects on vascular cells of cerebral arteries and arterioles. Hypoxia-induced drop in ATP levels opens K_{ATP} channels on smooth muscle, causing hyperpolarization and vasodilation [165]. In addition, hypoxia rapidly increases nitric oxide and adenosine production locally, also promoting vasodilation [166]. Chronic hypoxia increases cerebral blood flow through an effect on capillary density [16–19] (see *Microcirculation and Neurovascular Unit*).

EFFECT OF CARBON DIOXIDE

Carbon dioxide (CO_2) has a profound and reversible effect on cerebral blood flow, such that hypercapnia causes marked dilation of cerebral arteries and arterioles and increased blood flow, whereas hypocapnia causes constriction and decreased blood flow [167,168]. The potent vasodilator effect of CO_2 is demonstrated by the finding that in humans 5% CO_2 inhalation causes an increase in cerebral blood flow by 50% and 7% CO_2 inhalation causes a 100% increase in cerebral blood flow [168]. Although several mechanisms involved in hypercapnic vasodilation have been proposed, the major mechanism appears to be related to a direct effect of extracellular H^+ on vascular smooth muscle [169]. This is supported by findings that neither bicarbonate ion nor changes in Pco_2 alone affect cerebral artery diameter [170]. Other proposed mechanisms involved in the response to changes in Pco_2 include vasodilator prostanoids and nitric oxide; however, the involvement of these mediators appears to be species-specific [171,172].

CHAPTER 6

Barriers of the CNS

The first studies to demonstrate the existence of a selective barrier between the blood and the brain were done by Ehrlich in 1885 [173]. In a classic series of experiments, Ehrlich infused Evan's blue dye intravenously into a rat and found that all organs in the body stained except for the brain. However, he incorrectly surmised that the brain was made of tissue for which the dye could not adhere to. It was his graduate student, Goldmann, in 1913 that did the decisive experiment and injected dye into the CSF, finding that in this case, only the brain tissue stained [174]. He correctly surmised that there was a barrier between the brain and the blood. These experiments also determined that although there was a barrier between the blood and the brain, there was free access from CSF to brain and, therefore, there was no CSF–brain barrier (Figure 18). There are three main interfaces in the brain that protect neurons from blood-borne substances and help to maintain water homeostasis and an appropriate milieu for neuronal function: the *blood–CSF interface*, the *blood–brain interface (BBB)*, and the *CSF–blood interface* [175,176] (Figure 19). The cerebral endothelium forms the largest barrier in the brain, the BBB, while epithelial cells of the choroid plexus form the blood–CSF barrier, and the avascular arachnoid epithelium lies under the dura and completely encases the brain, forming the CSF–blood barrier. Other interfaces with blood and neural tissue include the blood–retinal barrier and the blood–spinal cord barrier. These barriers within the central nervous system provide several protective functions for the brain. They are protective against unwanted

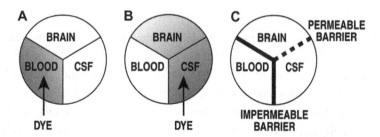

FIGURE 18: The blood–brain barrier, or BBB, to trypan blue and its diffusion from the cerebrospinal fluid, or CSF, into the brain. Used with permission from *Neuron* 2008;57:178–201.

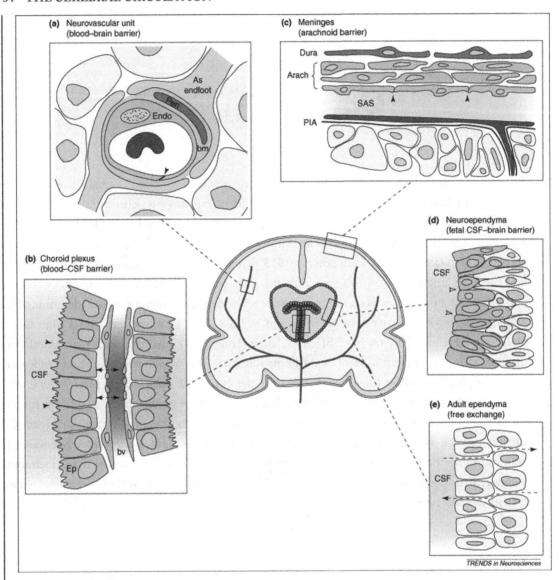

FIGURE 19: Schematics of the sites of the barrier interfaces (indicated in orange) in the adult and developing brain. (a) The blood–brain barrier is a barrier between the lumen of cerebral blood vessels and brain parenchyma. The endothelial cells (Endo) have luminal tight junctions (arrowhead) forming the physical barrier of the interendothelial cleft. Outside the endothelial cell is a basement membrane (bm) which also surrounds the pericytes (Peri). Around all these structures are the astrocytic endfeet processes from nearby astrocytes (As endfoot). All these structures together are often referred to as the neurovascular unit. (b) The blood–CSF barrier, a barrier between choroid plexus blood vessels and the CSF. The choroid plexus blood vessels are fenestrated and form a nonrestrictive barrier (small arrows); however, the epithelial cells (Ep) have apical tight junctions (arrowheads) that restrict intercellular pas-

pathogens and control the immunologic status of the brain [177]. The tight junctions at the BBB do not allow ions to move passively into the brain and thus prevent fluctuations in electrolytes that occur in the blood [178]. They also prevent proteins (albumin) and circulating blood cells (erythrocytes, leukocytes) from passing into the brain, which can damage neuronal tissue and interfere with tightly controlled water homeostasis [175–178].

THE BLOOD–CSF BARRIER

CSF is formed in the *lateral third* and *fourth ventricles* mainly by the *choroid plexus* and cerebral capillaries [179]. CSF functions as a cushion for the brain and spinal cord and provides important nutrients. CSF has the same composition as interstitial fluid (ISF) and mixes freely together across pial surfaces [179–181]. The capillaries of the choroid plexus do not have BBB properties, but are fenestrated and leaky [180]. However, the tight junctions of the ependymal cells of the choroid plexuses form the blood–CSF barrier [180]. Ependymal cells of the choroid plexus are epithelial-like. Ionic pumps, most importantly, the Na^+–K^+ ATPase on the apical surface of the ependymal cells produces the chemiosmotic energy for the osmotic gradient that contributes to fluid formation by the cells of the choroid plexus [182]. Water flow follows the osmotic gradient set up by the removal of three Na^+ ions for 2 K^+ ions. Formation of CSF by the choroid plexus is facilitated by the very high rates of blood flow to the choroid plexus [183].

CSF is formed at a rate of ~600 ml/day [184]. This rapid production results in turnover of CSF several times during the day. ISF formed by cerebral capillaries joins CSF formed by the choroid plexus in the cerebral ventricles and is the starting point of circulating CSF [181]. From the

sage of molecules. (c) The meningeal barrier is the least studied and structurally most complex of all the brain barriers. The blood vessels of the dura are fenestrated and provide little barrier function; however, the outer cells of the arachnoid membrane (Arach) have tight junctions (arrowheads), and this cell layer is believed to form the physical barrier between the CSF-filled subarachnoid space (SAS) and overlaying structures. The blood vessels in the arachnoid and on the pial surface (PIA) have tight junctions with similar barrier characteristics as cerebral blood vessels although lacking the surrounding pericytes and astrocytic endfeet. (d) The fetal CSF–brain barrier, a barrier between the CSF and brain parenchyma, has only been shown to be a functional barrier in the early developing. In early development, the neuro-ependymal cells are connected to each other by strap junctions (open arrowheads) that are believed to form the physical barrier restricting the passage of larger molecules such as proteins but not smaller molecules such as sucrose. (e) The adult ventricular ependyma. During development, the neuroependymal cells flatten and lose their strap junctions. The mature ependyma does not restrict the exchange of molecules at least as large as proteins between CSF and brain. Used with permission from *Trends Neurosci.* 2008 Jun;31(6):279-86.

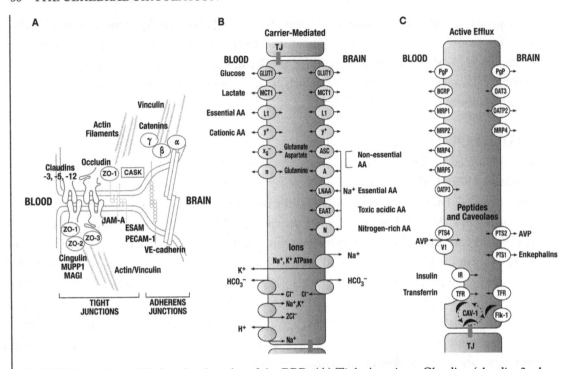

FIGURE 20: A simplified molecular atlas of the BBB. (A) Tight junctions. Claudins (claudin-3, claudin-5, and claudin-12) and occludin have four transmembrane domains with two extracellular loops. Zonula occludens proteins (ZO -1, ZO-2, and ZO-3) and the calcium-dependent serine protein kinase (CASK) are first-order cytoplasmic adaptor proteins that contain PDZ binding domains for the C terminus of the intramembrane proteins. Cingulin, multi-PDZ protein 1 (MUPP1), and the membrane-associated guanylate kinase with an inverted orientation of protein–protein interaction domain (MAGI) are examples of second-order adaptor molecules. The vascular endothelial cadherin (VE-cadherin) is the key molecule. Platelet endothelial cell adhesion molecule 1 (PECAM-1) mediates homophilic adhesion. Catenins (a, b, c) link adhesion junctions to actin/vinculin-based cytoskeleton. (B) Carrier-mediated transporters. GLUT1, glucose transporter, and monocarboxylate transporter 1 (MCT1) for lactate exist at both the luminal and the abluminal membranes. All essential amino acids (AA) are transported by the L1 and y+ systems on each membrane. Five Na^+-dependent transport systems mediate elimination of nonessential AA (ASC, A), essential AA (LNAA), the excitatory acidic AA (EAAT) (e.g., glutamate, aspartate), and nitrogen-rich AA (N) (e.g., glutamine) from the brain. Facilitative transporters xG and n on the luminal membrane mediate glutamate, aspartate, and glutamine efflux to blood. Ion transporters. The sodium pump (Na+, K+-ATPase) on the abluminal membrane controls Na+ influx and K+ efflux. Sodium–hydrogen exchanger on the luminal membrane is a key regulator of intracellular pH. Na+–K+–2Cl– cotransporter is on the luminal membrane. The chloride–bicarbonate exchanger exists on each membrane. (C) Active efflux transporters. Multidrug efflux transporters at the luminal membrane limit drug uptake into the brain. Transporters at the abluminal membrane could act in concert with luminal transporters to eliminate drugs from brain ISF. P-gp is expressed on each membrane. Breast cancer re-

ventricles, CSF exits into the cisterna magna and then over the cerebral covexities. CSF is absorbed at the arachnoid granulations into the blood before flowing up over the cerebral hemispheres.

THE BLOOD–BRAIN BARRIER

In the brain and spinal cord, the BBB is formed by cerebral endothelial cells that have highly specialized structural and functional properties [185,186]. Brain endothelial cells are phenotypically unique compared to endothelium in the periphery in that they have apical tight junction complexes that more closely resemble epithelium than endothelium [12,185,186]. In addition, while specialized tight junctions of the BBB limit passive diffusion of blood-borne solutes, brain endothelium also contains transporters that actively control transport of nutrients into the brain from the blood [12,185–187]. Like peripheral endothelium, cerebral endothelial cells are polarized and express specific transporters apically to actively transport nutrients from the blood into the brain and basolaterally to inactivate toxic substances and remove (efflux) them from the brain into the blood [12]. Thus, the cerebral endothelium provides a highly restricted, but controlled barrier to plasma constituents. Other unique features of the cerebral endothelium are a lack of fenestrations, a very low rate of pinocytosis that limits transcellular transport, and a high number of mitochondria associated with its high metabolic activity [6,7,12,88].

Although often overlooked, the BBB is present in cerebral endothelium throughout the brain, including pial arteries and arterioles and veins, but is absent from the circumventricular organs (CVO) [6,7,12,188]. The CVO are highly specialized areas in the brain and include area postrema and median eminence, neurohypophysis, pineal gland, sub-fornica organ, and lamina terminalis, which require significant cross-talk between the brain and peripheral blood, e.g., release and transport of hormones [188,189]. Therefore, the cerebral endothelium of the CVO are fenestrated and do not have BBB properties. It is often thought that the CVO are an area in the brain without barrier properties, but this is not the case. The barrier for the CVO lies in the epithelial cells known as tanycytes and ependymal cells. Thus, circulating substances can diffuse into the CVO but not beyond.

ULTRASTRUCTURE OF THE BBB

Ultrastructural studies by Reese and Karnovsky characterized brain endothelium as the morphologic site of the BBB [185]. Cerebral endothelial cells are connected paracellularly at junctional complexes by *tight junctions* and *adherens junctions* [12,188]. The molecular organization of the BBB tight junction and its adapter proteins that link to the actin cytoskeleton forms a continuous membrane that

---

sistance protein (BCRP) is on the luminal membrane. Multidrug resistance-associated proteins (MRPs) are expressed mainly on the luminal membrane. Organic anion transporting polypeptides (OATP) 2 and 3 exist on the luminal and abluminal membranes, respectively. Organic anion transporter 3 (OAT3) is on the abluminal membrane. Used with permission from *Neuron* 2008;57:178–201.

confers the high electrical resistance of the BBB (~1500–2000 Ω-cm^2) and retention of ions in the vascular lumen [12,99,178,188] (Figure 20). Many disease states including chronic disease, such as multiple sclerosis, experimental autoimmune encephalomyelitis, and Alzheimer's disease, and acute conditions, such as ischemic stroke, hypertension, and seizure, have been associated with dysregulation of tight junction proteins [12,99].

Tight Junctions

Tight junctions consist of three integral membrane proteins (claudin, occludin, and junction adhesion molecules (JAM)) and several accessory proteins including zona occludens (ZO) (ZO-1, ZO-2, ZO-3), cingulin, and others (190–193). Claudins are 22-kDa phosphoproteins that comprise the major component of tight junctions (194). Greater than 20 members of the claudin family have been identified [195]. They are located at the tight junction strand and bind other claudins homotypically on adjacent endothelial cells to form the primary seal of the tight junction [194]. The carboxy terminus of the claudins bind to cytoplasmic proteins including ZO-1, ZO-2, and ZO-3 (194). The zona occludens proteins (ZO-1, ZO-2, and ZO-3) together with cingulin and several others are cytoplasmic proteins involved in tight junction formation [196,197]. ZO-1 and ZO-2 bind to the actin cytoskeleton at their carboxy terminus [194]. This critical link provides for structural stability of the endothelial cell and is an important means of regulating paracellular permeability [99]. Occludin is a 65-kDa phosphoprotein with four transmembrane domains, a long carboxy-terminal cytoplasmic domain, and a short amino-terminal cytoplasmic domain [188,198,199]. Two extracellular loops of occludin and claudin originating in neighboring cells form the paracellular barrier of the tight junction [188]. Occludin is directly linked to the zona occludens proteins and thereby regulates permeability through their association with the actin cytoskeleton [200]. JAM are 40-kDa membrane proteins within the tight junction that also bind ZO-1 [201]. Of the three JAM molecules identified, only JAM-1 and JAM-3, but not JAM-2, are expressed in brain endothelium [202]. JAM-1 localizes with actin and is involved in cell-to-cell adhesion [202].

Adherens Junctions

Adherens junctions form adhesive contacts between cells and consist of the membrane protein cadherin that joins the actin cytoskeleton via intermediary proteins, the catenins [188,203]. Adherens junctions form from homophilic interactions between extracellular domains of cadherins on the surface of adjacent cells [204]. The cytoplasmic domains of cadherins bind to β- or γ-catenin, which are linked to the cytoskeleton via α-catenin [204]. Adherens junctions interact with tight junctions via ZO-1 and catenins to influence tight junction assembly [205].

REGULATION OF PARACELLULAR PERMEABILITY

The contractile activity of endothelial cells via actin stress fibers within the cytoplasm is of central importance to regulating tight junction permeability [99,206]. Agonists that promote relaxation of stress fibers (e.g., cyclic-AMP) decrease permeability by cell spreading, which strengthens cell–cell contact and reduces paracellular transport [207]. Alternatively, agonists that promote stress fiber contraction (e.g., PKC, VEGF) promote increased permeability by causing cell rounding, which decreases cell–cell contact [208]. Studies have demonstrated that inhibition of myosin light-chain (MLC) phosphorylation, which inhibits actin stress fiber contraction, decreases agonist-induced permeability [209,210]. Stress fiber activity as an underlying mechanism of paracellular transport is an important consideration to the numerous mediators of cerebral edema formation that are produced by the injured brain, e.g., histamine, bradykinin, arachidonic acid, etc. [99,206].

BBB TRANSPORTERS

The BBB freely passes oxygen, carbon dioxide, and small lipophilic substances, but is impermeable to hydrophilic molecules such as glucose, amino acids, and other nutrients essential to life [12]. Thus, a major physiological function of the BBB is the tight regulation of transport of nutrients and other molecules into and out of the brain. In addition, BBB transporters are involved in inactivation and reuptake of neurotransmitters [12,188]. The high electrical-resistant tight junctions only allow small lipid-soluble molecules (<400 Da) to cross the BBB [211]. All other substances must cross the BBB through specific transporters on either the apical or basolateral endothelial membrane [12,188]. Specific *carrier-mediated transport* (facilitated diffusion) systems facilitate transport of nutrients such as glucose and galactose, amino acids, nucleosides, purines, amines, and vitamins down their concentration gradient from the blood to the brain [12,188]. Transport of these nutrients is in general regulated by brain metabolic needs and the concentration of substrates in plasma. There are also *receptor-mediated transport* systems for proteins and peptides to transport neuroactive peptides, chemokines, and cytokines into the brain [2,188]. Large proteins such as transferrin, low-density lipoprotein (LDL), leptin, insulin, and insulin-like growth factor also use specific receptor-mediated transport systems to cross the BBB [264,265]. *Active efflux transporters* are located on both apical and basolateral endothelial membranes and are important for removal of molecules from the brain into the blood. A number of active efflux transporters have been identified, most belonging to a large family of proteins called the ATP-binding cassette (ABC) transporter superfamily [12,188]. Transporters such as these use ATP-bound energy for the transport of molecules across the cell membrane and include the multidrug resistance (MDR) transporter P-glycoprotein (P-gp) that mediates removal of toxic lipophilic metabolites and cationic drugs, multidrug resistance-associated proteins (MRP), the breast cancer resistance protein (BCRP), and other transporters of anionic compounds [12,188].

TRANSCELLULAR TRANSPORT

While tight junction proteins control paracellular movement of ions and solutes, a transcellular route also exists in cerebral endothelium for passage of lipophobic molecules through three distinct routes. *Fluid-phase endocytosis* is a constitutive process for passing macromolecules into the brain and for recycling the plasma membrane [212,213]. Molecules are internalized indiscriminantly without binding to the cell surface. In general, fluid-phase endocytosis is low in the cerebral endothelium, but is induced under pathologic conditions such as ischemic stroke and acute hypertension [206]. *Absorptive endocytosis* occurs when molecules such as lectins bind to carbohydrate moieties or to the negatively charged glycocaylx causing endocytosis [12,188]. *Receptor-mediated endocytosis* occurs when a specific molecule (ligand) binds to a receptor on the endothelial cell surface, triggering internalization of the receptor–ligand complex [12,188]. This process can be clathrin-mediated and involves cavaolae or clathrin-44-independent.

WATER HOMEOSTASIS IN THE BRAIN

The brain is unique in how it deals with water flow from the blood [178]. For all other tissues, there is convective water flow into the tissue with solute between endothelial cells such that only plasma proteins are retained in the vascular lumen [214]. Protein osmotic pressure offsets the efflux of fluid due to blood hydrostatic pressure and gives rise to Starling's forces. However, at the BBB, there is limited molecular transport due to a low rate of fluid-phase endocytosis, which limits transcellular flux, and coupling by high electrical resistance tight junctions, which limits paracellular flux [215,216). These morphologic features prevent the extravasation of large and small solutes and, importantly, ions [178,214,215]. The tight junctions of the brain effectively prevent the movement of hydrophilic substances, including plasma proteins and univalent cations such as Na^+ and K^+ [178]. These unique barrier properties modifies Starling's forces such that any movement of water into the brain by normal blood hydrostatic pressure (CPP) is immediately opposed by the osmotic pressure gradient set up by the ions retained in the vascular lumen [178]. This unique situation prevents vasogenic edema formation and is considered to be a protective role of the BBB.

HYDRAULIC CONDUCTIVITY

Another unique feature of cerebral endothelium is that is has an unusually high resistance to water filtration in response to hydrostatic pressure, a parameter known as *hydraulic conductivity* (Lp). Unlike measures of solute or tracer permeability, Lp is the critical transport parameter that relates water flux to hydrostatic pressure [178]. Together with transvascular filtration (Jv), Lp is an important determinant of the movement of water into the brain [217,218]. Lp is also a characteristic parameter of convective fluid motion and can influence the mass transport of solutes and other molecules through the endothelium [219]. Therefore, this parameter encompasses both transcellular and paracellular routes of permeability.

ROLE OF ASTROCYTES IN BBB FUNCTION

While it is the structural properties of the cerebral endothelium that make up the BBB (e.g., tight junctions), associated cells within the brain parenchyma contribute to its barrier properties, most notably astrocytes. Astrocytes are specific brain cells located among endothelium, pericytes, and neurons. An early concepts of the BBB was that astrocytic end-feet, which are in close proximity to cerebral endothelial cells, provided a structural barrier and thus contributed to BBB properties. However, studies by Reese and Karnovsky showed that the site of barrier function was at the level of the cerebral endothelial cells and not astrocytes [185]. A role for astrocytes in inducing BBB properties of the cerebral endothelium, as opposed to being a physical barrier, became apparent [220]. *In vitro* cell culture studies confirm an important role of glial cells in inducing BBB phenotype, including upregulation of tight junction proteins [221,222]. The interaction of astrocytes and cerebral endothelium is not one way. There is considerable evidence that the cerebral endothelium signals astrocytes. For example, the water channel aquaporin-4 (AQP-4) is primarily expressed only in astrocytic end-feet surrounding vessels in the brain parenchyma, but not in astrocytes just interacting with neurons [223] (Figure 21). Thus, the maintenance of BBB properties and function likely

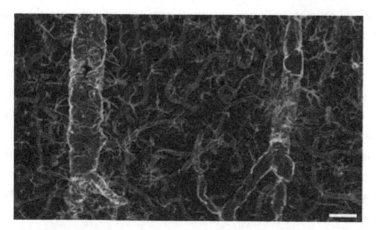

FIGURE 21: Double immunolabeling of AQP-4 (red) and GFAP (green). AQP-4 immunolabeling reveals that the entire network of vessels, including capillaries, is covered by astrocytic processes, albeit GFAP-negative. Smaller vessels and capillaries are mostly GFAP-negative but display intense labeling against the astrocyte-specific channel AQP-4. The AQP-4 labeling reveals continuous coverage by astrocytic end-feet. Scale bar = 60 μm. Used with permission from *J Neurosci* 2003;23:9254–9262.

depends on cross-talk between the endothelium and astrocytes. In addition, astrocytes have a large number of K$^+$ channels (K$_{ir}$4.1 and rSloK$_{Ca}$) and spatially buffer K$^+$ in the perivascular space [224]. Astrocytes have other important functions in regulating water and ionic homeostasis in the brain and are important contributors to cytotoxic edema in the brain during injury [206] (see *Vasogenic Edema*).

CEREBRAL EDEMA FORMATION

Cytotoxic vs. Vasogenic Edema

Klatzo first characterized brain edema as cytotoxic vs. vasogenic depending on whether or not the BBB is disrupted [225]. Cytotoxic edema occurs when brain cells swell at the expense of the extracellular space, but BBB properties are present. Vasogenic edema occurs when there is increased permeability of the BBB, which allows an influx of plasma constituents, and water, which expand the extracellular space. Several pathological conditions cause breakdown of the BBB and vasogenic edema, including ischemic stroke, acute hypertension, seizure, and traumatic brain injury. Cell swelling associated with cytotoxic edema may be compensatory or even protective. Precapillary astrocyte end-feet are the first cellular elements to swell during ischemia [226], a process thought to normalize the composition of the extracellular environment for normal neuronal activity [227]. Glial cells can also inactivate neurotransmitter [228], take up excess potassium ions produced during neuronal activity [229], and scavenge reactive oxygen species [230] (see *Role of Astrocytes*).

Under normal conditions, the barrier property of the cerebral endothelium has low hydraulic conductivity, and prevents bulk flow of water, ions, and proteins into the brain from hydrostatic forces such as blood pressure [231]. When the BBB is disrupted (e.g., during ischemia or acute hypertension), hydrostatic forces become significant enough that the rate of protein entry into the brain is directly related to the pressure gradient between the blood and the brain. However, while the passage of protein from the plasma into the brain is a measure of BBB disruption, it does not significantly contribute to edema formation. Albumin and other proteins passing into the brain have been used as a measure of BBB disruption; however, the concentration of a large protein is several orders of magnitude smaller than that of ions. Therefore, the increase in osmolality that occurs when albumin enters the brain is small compared to that of ions. Albumin and protein entry into the brain may be an indicator of BBB disruption, but its contribution to edema formation is small compared to that of ions.

Summary

The brain's circulation has many unique properties that under normal conditions help maintain constant perfusion to an organ with high metabolic demands. In addition, the unique structural properties of the cerebral endothelium, including tight junctions that do not allow ions to pass freely, a low rate of transcellular transport, and very low hydraulic conductivity, help to maintain an appropriate ionic milieu for neuronal function and to control water homeostasis in an organ that has limited capacity to expand within the skull. When disrupted by trauma or disease, these important features of the cerebral circulation are lost and brain injury can occur. Although a great deal has been learned regarding the structure and function of the cerebral circulation over the past few decades, the use of transgenic animals and sophisticated imaging has and will continue to provide greater insight into cerebrovascular function and how it can be preserved to prevent brain injury.

References

[1] Jones EG. On the mode of entry of blood vessels into the cerebral cortex. *J Anat.* 1970; 106: pp. 507–520.

[2] Rennels M, Nelson E. Capillary innervation in the mammalian central nervous system: an electron microscope demonstration (1). *Am J Anat.* 1975; 144: pp. 233–241.

[3] Cohen Z, Bonvento G, Lacombe P, Hamel E. Serotonin in the regulation of brain microcirculation. *Prog Neurobiol.* 1996; 50: pp. 335–362.

[4] Cipolla MJ, Li Rui, Vitullo L. Perivascular innervation of penetrating brain parenchymal arterioles. *J Cardiovasc Pharm.* 2004; 44(1): pp. 1–8.

[5] Nishimura N, Schaffer CB, Friedman B, Lyden PD, Kleinfeld D. Penetrating arterioles are the bottleneck in the perfusion of neocortex. *Proc Natl Acad Sci USA.* 2007; 104: pp. 365–370.

[6] Roggendorf W, Cervos-Navarro J. Ultrastructure of arterioles in the cat brain. *Cell Tissue Res.* 1977; 178: pp. 495–515.

[7] Abbott NJ. Inflammatory mediators and modulation of blood–brain barrier permeability. *Cell Mol Neurobiol.* 2000; 2: pp. 131–147.

[8] Schaller B. Physiology of cerebral venous blood flow: from experimental data in animals to normal function in humans. *Brain Res Brain Res Rev.* 2004; 46: pp. 243–260.

[9] Kiliç T, Akakin A. Anatomy of cerebral veins and sinuses. *Frontiers Neurol Neurosci.* 2008; 23: pp. 4–15.

[10] Lee RM. Morphology of cerebral arteries. *Pharmacol Ther.* 1995;66: pp. 149–173.

[11] Begley DJ, and Brightman MW. Structural and functional aspects of the blood–brain barrier. *Prog Drug Res.* 2003;61: pp. 39–78.

[12] Zlokovic BV. The blood–brain barrier in health and chronic neurodegenerative disorders. *Neuron.* 2008; 57: pp. 178–201.

[13] Zlokovic BV. Neurovascular mechanisms of Alzheimer's neurodegeneration. *Trends Neurosci.* 2005; 28: pp. 202–208.

[14] Wei L, Otsuka T, Acuff V, Bereczki D, Pettigrew K, Patlak C, Fenstermacher J. The velocities of red cell and plasma flows through parenchymal microvessels of rat brain are decreased by pentobarbital. *J Cereb Blood Flow Metab.* 1993; 13: pp. 487–497.

[15]   Klein B, Kuschinsky W, Schrock H, Vetterlein F. Interdependency of local capillary density, blood flow, and metabolism in rat brains. *Am J Physiol*. 1986; 251: pp. H1333–H1340.

[16]   Xu K, Lamanna JC. Chronic hypoxia and the cerebral circulation. *J Appl Physiol*. 2006 100: pp. 725–730.

[17]   Boero JA, Ascher J, Arregui A, Rovainen C, Woolsey TA. Increased brain capillaries in chronic hypoxia. *J Appl Physiol*. 1999; 86: pp. 1211–1219.

[18]   Dunn JF, Roche MA, Springett R, Abajian M, Merlis J, Daghlian CP, Lu SY, and Makki M. Monitoring angiogenesis in brain using steadystate quantification of DeltaR2 with MION infusion. *Magn Reson Med*. 2004; 51: pp. 55–61.

[19]   Dunn JF, Grinberg O, Roche M, Nwaigwe CI, Hou HG, and Swartz HM. Noninvasive assessment of cerebral oxygenation during acclimation to hypobaric hypoxia. *J Cereb Blood Flow Metab*. 2000; 20: pp. 1632–1635.

[20]   Sokolova IA, Manukhina EB, Blinkov SM, Koshelev VB, Pinelis VG, Rodionov IM. Rarefaction of the arterioles and capillary network in the brain of rats with different forms of hypertension. *Microvasc Res*. 1985; 30: pp. 1–9.

[21]   Ballabh P, Braun A, Nedergaard M. The blood–brain barrier: an overview: structure, regulation, and clinical implications. *Neurobiol Dis*. 2004; 16: pp. 1–13.

[22]   Hamel E. Perivascular nerves and the regulation of cerebrovascular tone. *J Appl Physiol*. 2006; 100: pp. 1059–1064.

[23]   Lok J, Gupta P, Guo S, Kim WJ, Whalen MJ, van Leyen K, Lo EH. Cell–cell signaling in the neurovascular unit. *Neurochem Res*. 2007; 32: pp. 2032–2045.

[24]   Dore-Duffy P. Pericytes: pluripotent cells of the blood brain barrier. *Curr Pharm Des*. 2008; 14: pp. 1581–1593.

[25]   Allt G, Lawrenson JG. Pericytes: cell biology and pathology. *Cells Tissues Organs*. 2001; 169: pp. 1–11.

[26]   Dore-Duffy P, La Manna JC. Physiologic angiodynamics in the brain. *Antioxid Redox Signal*. 2007; 9: pp. 1363–1372.

[27]   Hossmann KA. Pathophysiology and therapy of experimental stroke. *Cell Mol Neurobiol*. 2006; 26: pp. 1057–1083.

[28]   Liebeskind DS. Collateral circulation. *Stroke*. 2003; 34: pp. 2279–2284.

[29]   Handa Y, Caner H, Hayashi M, Tamamaki N, Nojyo Y. The distribution pattern of the sympathetic nerve fibers to the cerebral arterial system in rat as revealed by antegrade labeling with WGA-HRP. *Exp Brain Res*. 1990; 82: pp. 493–498.

[30]   Cohen Z, Bovento G, Lacombe P, Seylaz J, MacKenzie ET, Hamel E. Cerebrovascular nerve fibers immunoreactive for tryptophan-5-hydroxylase in the rat: distribution, putative origin and comparison with sympathetic noradrenergic nerves. *Brain Res*. 1992; 598: pp. 203–214.

[31] Chédotal A, Hamel E. Serotonin-synthesizing nerve fibers in rat and cat cerebral arteries and arterioles: immunohistochemistry of tryptophan-5-hydroxylase. *Neurosci Lett*. 1990; 116: pp. 269–274.

[32] Cohen Z, Molinatti G, Hamel E. Astroglial and vascular interactions of noradrenaline terminals in the rat cerebral cortex. *J Cereb Blood Flow Metab*. 1997; 17: pp. 894–904.

[33] Bleys RLAW, Cowen T. Innervation of cerebral blood vessels: morphology, plasticity, age-related, and Alzheimer's disease-related neurodegeneration. *Microsc Res Tech*. 2001; 53: pp. 106–188.

[34] Estrada C, Mengual E, González C. Local NADPH-diaphorase neurons innervate pial arteries and lie close or project to intracerebral blood vessels: a possible role for nitric oxide in the regulation of cerebral blood flow. *J Cereb Blood Flow Metab*. 1993; 13: pp. 978–984.

[35] Allaman I, Pellerin L, Magistretti PJ. Protein targeting to glycogen mRNA expression is stimulated by noradrenaline in mouse cortical astrocytes. *Glia*. 2000; 30: pp. 382–391.

[36] Kotter K, Klein J. Adrenergic modulation of astroglial phospholipase D activity and cell proliferation. *Brain Res*. 1999; 29: pp. 138–145.

[37] Cohen Z, Bouchelet I, Olivier A, Villemure JG, Ball R, Stanimirovic DB, Hamel E. Multiple microvascular and astroglial 5-hydroxytryptamine receptor subtypes in human brain: molecular and pharmacologic characterization. *J Cereb Blood Flow Metab*. 1999; 19: pp. 908–917.

[38] Xu T, Pandey SC. Cellular localization of serotonin(2A) (5HT(2A)) receptors in the rat brain. *Brain Res Bull*. 2000; 51: pp. 499–505.

[39] Sandén N, Thorlin T, Blomstrand F, Persson PA, Hansson E. 5-Hydroxytryptamine2b receptors stimulate Ca2+ increases in cultured astrocytes from three different brain regions. *Neurochem Int*. 2000; 36: pp. 427–434.

[40] Sándor P. Nervous control of the cerebrovascular system: doubts and facts. *Neurochem Int*. 1999; 35: pp. 237–259.

[41] Lincoln J. Innervation of cerebral arteries by nerves containing 5-hydroxytryptamine and noradrenaline. *Pharmacol Ther*. 1995;68: pp. 473–501.

[42] Högestatt ED, Andersson KE. On the postjunctional α-adrenoreceptors in rat cerebral and mesenteric arteries. *J Anat Pharmacol*. 1984; 4: pp. 161–175.

[43] Alberts B, Johnson A, Lewis J, et al. Signaling through G-protein-linked cell-surface receptors. In: *Molecular Biology of the Cell*, Alberts B, Johnson A, Lewis J, et al. (Eds.). New York: Garland Science, 2002; pp. 852–862.

[44] Dacey RG, Duling BR. Effect of norepinephrine on penetrating arterioles of rat cerebralcortex. *Am J Physiol*. 1984; 246: pp. H380–H385.

[45] Rosendorff C, Mitchell G, Mitchell D. Adrenergic innervation affecting local cerebral blood flow. In: *Neurogenic Control of Brain Circulation: Werner-Gren Center International*

Symposium Series, Vol. 30, Owman Ch, Edvinsson L (Eds.), Oxford: Pergamon, 1977; pp. 455–464.

[46]   Sercombe R, Hardebo JE, Kåhrström J, Seylaz J. Amine-induced responses of pial and penetrating cerebral arteries: evidence for heterogeneous responses. *J Cereb Blood Flow Metab.* 1990; 10: pp. 808–818.

[47]   Mayhan WG. Responses of cerebral arterioles to activation of β-adrenergic receptors during diabetes mellitus. *Stroke.* 1994; 25: pp. 141–146.

[48]   Harper AM, MacKenzie ET. Effects of 5-hydroxytryptamine on pial arteriolar calibre in anaesthetized cat. *J Physiol.* 1977; 271: pp. 735–746.

[49]   Mayhan WG, Faraci FM, and Heistad DD. Responses of cerebral arterioles to adenosine, 5'-diphosphate, serotonin, and the thromboxane analog U-46619 during chronic hypertension. *Hypertension.* 1988; 12: pp. 556–561.

[50]   Nilsson T, Longmore J, Shaw D, Olesen IJ, Edvinsson L. Contractile 5-HT1B receptors in human cerebral arteries: pharmacologic characterization and localization with immunocytochemistry. *Br J Pharmacol.* 1999; 128: pp. 1133–1140.

[51]   Willis T. Cerebi Anatome, cui accessit nervorum, descriptio et usus. *Flesher J.* 1664.

[52]   Bleys RLAW, Cowen T. Innervation of cerebral blood vessels: morphology, plasticity, age-related, and Alzheimer's disease-related neurodegeneration. *Microsc Res Tech.* 2001; 53: pp. 106–188.

[53]   Kobayashi S, Tsukahara S, Sugita K, et al. Adrenergic and cholinergic innervations of rat cerebral arteries. Consecutive demonstration on whole mount preparations. *Histochemistry.* 1981; 70: pp. 129–138.

[54]   Handa Y, Caner H, Hayashi M, et al. The distribution pattern of the sympathetic nerve fibers to the cerebral arterial system in rat as revealed by antegrade labeling with WGA-HRP. *Exp Brain Res.* 1990; 82: pp. 493–498.

[55]   Hamel E, Edvinsson L, McKenzie ET. Heterogeneous vasomotor responses of anatomically distinct feline cerebral arteries. *Br J Pharmacol.* 1988; 94: pp. 423–436.

[56]   Edvinsson L, Egund N, Owman C, Sahlin C, Svendgaard NA. Reduced noradrenaline uptake and retention in cerebrovascular nerves associated with angiographically visible vasoconstriction following experimental subarachnoid hemorrhage. *Brain Res Bull.* 1982; 9: pp. 799–805.

[57]   Wahl M. Local chemical, neural, and humoral regulation of cerebrovascular resistance vessels. *J Cardiovasc Pharm.* 1985; 7(Suppl 3): pp. S36–S46.

[58]   Faraci FM, Heistad DD. Regulation of large cerebral arteries and cerebral microvascular pressure. *Circ Res.* 1990; 66: pp. 8–17.

[59]   Tuor UI. Acute hypertension and sympathetic stimulation: local heterogeneous changes in cerebral blood flow. *Am J Physiol.* 1992; 263(2 Pt 2): pp. H511–H518.

[60]   Goadsby PJ, Edvinsson L. Neurovascular control of the cerebral circulation. In: *Cerebral Blood Flow and Metabolism*, 2nd ed., Edvinsson L, Krause DN (Eds.). Philadelphia, PA: Lippincott Williams & Wilkins, 2002, pp. 172–188.

[61]   Suzuki N, Hardebo JE. The cerebrovascular parasympathetic innervation. *Cerebrovasc Brain Metab Rev.* 1993; 5: pp. 33–46.

[62]   Talman WT, Nitschke Dragon D. Neuronal nitric oxide mediates cerebral vasodilatation during acute hypertension. *Brain Res.* 2007; 1139: pp. 126–132.

[63]   Asahi M, Huang Z, Thomas S, Yoshimura S, Sumii T, Mori T, Qiu J, Amin-Hanjani S, Huang PL, Liao JK, Lo EH, Moskowitz MA. Protective effects of statins involving both eNOS and tPA in focal cerebral ischemia. *J Cereb Blood Flow Metab.* 2005; 25: pp. 722–729.

[64]   Waeber C, Moskowitz MA. Migraine as an inflammatory disorder. *Neurology.* 2005; 64: pp. S9–S15.

[65]   Bolay H, Reuter U, Dunn AK, Huang Z, Boas DA, and Moskowitz MA. Intrinsic brain activity triggers trigeminal meningeal afferents in a migraine model. *Nat Med.* 2002; 8: pp. 136–142.

[66]   Edvinsson L, Uddman R, Juul R. Peptidergic innervation of the cerebral circulation. Role in subarachnoid hemorrhage in man. *Neurosurg Rev.* 1990; 13: pp. 265–272.

[67]   Bayliss N. On the local reactions of the arterial wall to changes of internal pressure. *J Physiol.* 1902; 28: pp. 220–231.

[68]   Kontos HA, Wei EP, Raper AJ, Rosenblum WI, Navari RM, Patterson JL Jr. Role of tissue hypoxia in local regulation of cerebral microcirculation. *Am J Physiol.* 1978; 234: pp. H582–H591.

[69]   Mellander S. Functional aspects of myogenic vascular control. *J Hypertens.* 1989; 7:(Suppl 4): pp. S21–S30.

[70]   Busija DW, Heistad DD. Factors involved in the physiological regulation of the cerebral circulation. *Rev Physiol Biochem Pharamacol.* 1984; 101: pp. 161–211.

[71]   Osol G, Brekke JF, McElroy-Yaggy K, Gokina NI. Myogenic tone, reactivity, and forced dilatation: a three-phase model of in vitro arterial myogenic behavior. *Am J Physiol Heart Circ Physiol.* 2002; 283: pp. H2260–2267.

[72]   Cipolla MJ, Osol G. Vascular smooth muscle actin cytoskeleton in cerebral artery forced dilatation. *Stroke.* 1998; 29: pp. 1223–1228.

[73]   Paternò R, Heistad DD, Faraci FM. Potassium channels modulate cerebral autoregulation during acute hypertension. *Am J Physiol Heart Circ Physiol.* 2000; 278: pp. H2003–2007.

[74]   Schubert R, Lidington D, Bolz SS. The emerging role of $Ca2+$ sensitivity regulation in promoting myogenic vasoconstriction. *Cardiovasc Res.* 2008; 77: pp. 8–18.

[75]   Knot HJ, Nelson MT. Regulation of arterial diameter and wall [Ca2+] in cerebral arteries of rat by membrane potential and intravascular pressure. *J Physiol.* 1998; 508: pp. 199–210.

[76]   Moosmang S, Schulla V, Welling A, Feil R, Feil S, Wegener JW et al. Dominant role of smooth muscle L-type calcium channel Cav1.2 for blood pressure regulation. *EMBO J.* 2003; 22: pp. 6027–6034.

[77]   Johnson PC. The myogenic response in the microcirculation and its interaction with other control systems. *J Hypertens.* 1989; 7(Suppl 4): pp. S33–S39.

[78]   Hui Z, Ratz PH, Hill MA. Role of myosin phosphorylation and Ca in myogenic reactivity and arteriolar tone. *Am J Physiol.* 1995; 269: pp. H1590–H1596.

[79]   Welsh DG, Morielli AD, Nelson MT, Brayden JE. Transient receptor potential channels regulate myogenic tone of resistance arteries. *Circ Res.* 2002; 90: pp. 248–250.

[80]   Earley S, Waldron BJ, Brayden JE. Critical role for transient receptor potential channel TRPM4 in myogenic constriction of cerebral arteries. *Circ Res.* 2004; 95: pp. 922–929.

[81]   Dopico AM, Kirber MT, Singer JJ, Walsh JV. Membrane stretch directly activates large conductance Ca-activated calcium channels in mesenteric artery smooth muscle cells. *Am J Hypertens.* 1994; 7: pp. 82–89.

[82]   Nelson MT, Conway MA, Knot HJ, Brayden JE. Chloride channel blockers inhibit myogenic tone in rat cerebral arteries. *J Physiol.* 1997; 502: pp. 259–264.

[83]   D'Angelo G, Mogford JE, Davis GE, Davis MJ, Meininger GA. Integrin-mediated reduction in vascular smooth muscle Ca induced by RDG-containing peptide. *Am J Physiol.* 1997; 272: pp. H2065–H2070.

[84]   Cipolla MJ, Gokina NI, Osol G. Pressure-induced actin polymerization in vascular smooth muscle as a mechanism underlying myogenic behavior. *FASEB J.* 2002; 16: pp. 72–76.

[85]   Geiger B, Spatz JP, Bershadsky AD. Environmental sensing through focal adhesions. *Nat Rev Mol Cell Biol.* 2009; 10: pp. 21–33.

[86]   Clark K, Middelbeek J, van Leeuwen FN. Interplay between TRP channels and the cytoskeleton in health and disease. *Eur J Cell Biol.* 2008; 87: pp. 631–640.

[87]   Coulson RJ, Cipolla MJ, Vitullo L, Chesler NC. Mechanical properties of rat middle cerebral arteries with and without myogenic tone. *J Biomed Eng.* 2004; 126: pp. 76–81.

[88]   Schubert R, Kalentchuk VU, Krien U. Rho kinase inhibition partly weakens myogenic reactivity in rat small arteries by changing calcium sensitivity. *Am J Physiol Heart Circ Physiol.* 2002; 283: pp. H2288–H2295.

[89]   Lagaud G, Gaudreault N, Moore ED, van Breemen C, Laher I. Pressure dependent myogenic constriction of cerebral arteries occurs independently of voltage-dependent activation. *Am J Physiol Heart Circ Physiol.* 2002; 283: pp. H2187–H2195.

[90]   Osol G, Laher I, Cipolla MJ. Protein kinase C modulates basal myogenic tone in resistance arteries from the cerebral circulation. *Circ Res.* 1991; 68: pp. 359–367.

[91]   Jaggar JH, Porter VA, Lederer WJ, Nelson MT. Calcium sparks in smooth muscle. *Am J Physiol Cell Physiol*. 2000; 278: pp. C235–C256.

[92]   Jaggar JH. Intravascular pressure regulates local and global Ca(2+) signaling in cerebral artery smooth muscle cells. *Am J Physiol Cell Physiol*. 2001; 281: pp. C439–C448.

[93]   Brayden JE, Nelson MT. Regulation of arterial tone by activation of calcium-dependent potassium channels. *Science*. 1992; 256: pp. 532–535.

[94]   Iadecola C, Yang G, Ebner TJ, Chen G. Local and propagated vascular responses evoked by focal synaptic activity in cerebellar cortex. *J Neurophysiol*. 1997; 78: pp. 651–659.

[95]   Cipolla MJ, McCall A, Lessov N, and Porter J. Reperfusion decreases myogenic reactivity and alters middle cerebral artery function after focal cerebral ischemia in rats. *Stroke*. 1997; 28: pp. 176–180.

[96]   Dirnagl U, Iadecola C, Moskowitz MA. Pathobiology of ischaemic stroke: an integrated view. *Trends Neurosci*. 1999; 22: pp. 391–397.

[97]   Shima K. Hydrostatic brain edema: basic mechanisms and clinical aspect. *Acta Neurochir*. 2003; 86(Suppl): pp. 17–20.

[98]   Iadecola C. Neurovascular regulation in the normal brain and in Alzheimer's disease. *Nat Rev Neurosci*. 2004; 5: pp. 347–360.

[99]   Hawkins BT, Davis TP. The blood–brain barrier/neurovascular unit in health and disease. *Pharmacol Rev*. 2005; 57: pp. 173–185.

[100]   Oby E, Janigro D. The blood–brain barrier and epilepsy. *Epilepsia*. 2006; 47: pp. 1761–1774.

[101]   Faraci FM, Heistad DD. Regulation of the cerebral circulation: role of endothelium and potassium channels. *Physiol Rev*. 1998; 78: pp. 53–97.

[102]   Faraci FM, Brian JE. Nitric oxide and the cerebral circulation. *Stroke*. 1994; 25: pp. 692–703.

[103]   Moncada S, Palmer RMJ, Higgs EA. Nitric oxide: physiology, pathophysiology and pharmacology. *Pharmacol Rev*. 1992; 43: pp. 109–142.

[104]   Szabo C. Physiologic and pathological roles of nitric oxide in the central nervous system. *Brain Res Bull*. 1991; 41: pp. 131–141.

[105]   Cipolla MJ, Smith J, Kohlmeyer MM, Godfrey JA. SKCa and IKCa Channels, myogenic tone, and vasodilator responses in middle cerebral arteries and parenchymal arterioles: effect of ischemia and reperfusion. *Stroke*. 2009; 40: pp. 1451–1457.

[106]   Sobey CG, Faraci FM. Effects of a novel inhibitor of guanylyl cyclase on dilator responses of mouse cerebral arterioles. *Stroke*. 1997; 28: pp. 837–842.

[107]   Robertson BE, Schubert R, Hescheler J, Nelson MT. cGMP-dependent protein kinase activates Ca-activated K channels in cerebral artery smooth muscle. *Am J Physiol*. 1993; 265: pp. C299–C303.

[108]   Tayeh MA, Marletta MA. Macrophage oxidation of L-arginine to nitric oxide, nitrite, and nitrate. Tetrahydrobiopterin is required cofactor. *J Biol Chem.* 1989; 264: pp. 19654–19658.

[109]   Xia Y, Tsai AL, Berka V, Zweier JL. Superoxide generation from endothelial nitric oxide synthase. A Ca2+/calmodulin-dependent and tetrahydrobiopterin regulatory process. *J Biol Chem.* 1998; 273: pp. 25804–25808.

[110]   Katusic Z. Vascular endothelial dysfunction: does tetrahydrobiopterin play a role? *Am J Physiol.* 2001; 281: pp. H981–H986.

[111]   Fukai T. Endothelial GTPCH in eNOS uncoupling and atherosclerosis. *Arterioscler Thromb Vasc Biol.* 2007; 27: pp. 1493–1495.

[112]   Landmesser U, Dikalov S, Price SR, McCann L, Fukai T, Holland SM, Mitch WE, Harrison DG. Oxidation of tetrahydrobiopterin leads to uncoupling of endothelial cell nitric oxide synthase in hypertension. *J Clin Invest.* 2003; 111: pp. 1201–1209.

[113]   Pannirselvam M, Simon V, Verma S, Anderson T, Triggle CR. Chronic oral supplementation with sepiapterin prevents endothelial dysfunction and oxidative stress in small mesenteric arteries from diabetic (db/db) mice. *Br J Pharmacol.* 2003; 140: pp. 701–706.

[114]   Bauer PM, Fulton D, Boo YC, Sorescu GP, Kemp BE, Jo H, Sessa WC. Compensatory phosphorylation and protein–protein interactions revealed by loss of function and gain of function mutants of multiple serine phosphorylation sites in endothelial nitric-oxide synthase. *J Biol Chem.* 2003; 278: pp. 14841–14849.

[115]   Dudzinski DM, Michel T. Life history of eNOS: partners and pathways. *Cardiovasc Res.* 2007; 75: pp. 247–260.

[116]   Marrelli SP, Eckmann MS, Hunte MS. Role of endothelial intermediate conductance Kca channels in cerebral EDHF-mediated dilations. *Am J Physiol.* 2003; 285: pp. H1590–H1599.

[117]   McNeish AJ, Sandow SL, Neylon CB, Chen MX, Dora KA, Garland CJ. Evidence for involvement of both IKCa and SKCa channels in hyperpolarizing responses of the rat middle cerebral artery. *Stroke.* 2006; 37: pp. 1277–1282.

[118]   Busse R, Edwards G, Feletou M, Fleming I, Vanhoutte PM, Weston AH. EDHF: bringing the concepts together. Trends Pharmacol Sci.2003; 23: pp. 374–380.

[119]   Griffith TM. Endothelium-dependent smooth muscle hyperpolarization: do gap junctions provide a unifying hypothesis? *Br J Pharmacol.* 2004; 141: pp. 881–903.

[120]   Edwards G, Dora KA, Gardener MJ, Garland CJ, Weston AH. K+ is an endothelium-dependent hyperpolarizing factor in rat arteries. *Nature.* 2998; 396: pp. 269–272

[121]   Miura H, Guttermann DD. Human coronary arteriolar dilation to arachidonic acid depends on cytochrome P450 monooxygenase and Ca2+-activated K+-channels. *Circ Res.* 1998; 83(5): pp. 501–507.

[122]   Lacza Z, Puskar M, Kis B, Perciaccante JV, Miller AW, Busija DW. Hydrogen peroxide acts

as an EDHF in the piglet pial vasculature in response to bradykinin. *Am J Physiol.* 2002; 283: pp. H406–H411.

[123] Si H, Heyken W-T, Wolfle SE, Tysiac M, Schubert R, Grgic I, Vilianovich L, Gieging G, Maier T, Gross V, Bader M, DeWit C, Hoyer J, Kohler R. Impaired endothelium-derived hyperpolarizing factor-mediated dilations and increased blood pressure in mice deficient of the intermediate-conductance Ca2+-activated K+ channel. *Circ Res.* 2006; 99: pp. 537–544.

[124] Kohler R, Degenhardt C, Kuhn M, Funkel N, Paul M, Hoyer J. Expression and function of endothelial ca(2+)-activated K(+) channels in human mesenteric artery: single-cell reverse transcriptase–polymerase chain reaction and electrophysiological study in situ. Cir Res.2000; 87: pp. 496–503.

[125] Sandow SL, Haddock RE, Hill CE, Chadha PS, Kerr PM, Welsh DG, Plane F. What's where and why at a vascular myoendothelial microdomain signalling complex. *Clin Exp Pharmacol Physiol.* 2009; 36: pp. 67–76.

[126] Sandow SL. Factors, fiction and endothelium-derived hyperpolarizing factor. *Clin Exp Pharmacol Physiol.* 2004; 31: pp. 563–570.

[127] Andresen J, Shafi NI, Bryan RM Jr. Endothelial influences on cerebrovascular tone. *J Appl Physiol.* 2006; 100: pp. 318–327.

[128] Bogatcheva NV, Sergeeva MG, Dudek SM, and Verin AD. Arachidonic acid cascade in endothelial pathobiology. *Microvasc Res.* 2005; 69: pp. 107–127.

[129] Smith WL, Garavito RM, and DeWitt DL. Prostaglandin endoperoxide H synthases (cyclooxygenases)-1 and -2. *J Biol Chem.* 1996; 271: pp. 33157–33160.

[130] DeWitt DS, Kong DL, Lyeth BG, Jenkins LW, Hayes RL, Wooten ED, and Prough DS. Experimental traumatic brain injury elevates brain prostaglandin E2 and thromboxane B2 levels in rats. *J Neurotrauma.* 1988; 5: pp. 303–313.

[131] Cole DJ, Patel PM, Schell RM, Drummond JC, and Osborne TN. Brain eicosanoid levels during temporary focal cerebral ischemia in rats: a microdialysis study. *J Neurosurg Anesthesiol.* 1993; 5: 41–47.

[132] You JM, Golding EM, and Bryan RM. Arachidonic acid metabolites, hydrogen peroxide, and EDHF in cerebral arteries. *Am J Physiol Heart Circ Physiol.* 2005; 289: pp. H1077–H1083.

[133] Clarke DD, Sokoloff L. Circulation and energy metabolism of the brain. In: *Basic Neurochemistry,* Siegel G, Agrano BV, Albers RW, Molino PV (Eds.). New York: Raven Press, 1989: pp. 565–590.

[134] Drake CT, Iadecola C. The role of neuronal signaling in controlling cerebral blood flow. *Brain Lang.* 2007; 102: pp. 141–152.

[135] Kulik T, Kusano Y, Aronhime S, Sandler AL, Winn HR. Regulation of cerebral vasculature in normal and ischemic brain. *Neuropharmacology.* 2008; 55: pp. 281–288.

[136]   Ku DN, Zhu C. In: *Hemodynamic Forces and Vascular Cell Biology*, Sumpio BE (Ed.), Austin, TX: CRC Press, RG Landes Co., 1993: p. 3.

[137]   Paulson OB, Strandgaard S, Edvinsson L. Cerebral autoregulation. *Cerebrovasc Brain Metab Rev.* 1990; 2: pp. 161–192.

[138]   Phillips SJ, Whisnant JP. Hypertension and the brain. *Arch Intern Med.* 1992; 152: pp. 938–945.

[139]   Heistad DD, Kontos HA. In: *Handbook of Physiology: The Cardiovascular System III*, Berne RM, Sperelakis N (Eds.). Bethesda, MD: American Physiological Society, 1979: pp. 137–182 .

[140]   Hossmann K-A. Viability thresholds and the penumbra of focal ischemia. *Ann Neurol.* 1994; 36: pp. 557–565.

[141]   Iadecola C. Cerebral circulatory dysregulation in ischemia. In *Cerebrovascular Diseases*, Ginsberg MD, Bogousslavsky J. (Eds.). Cambridge, MA: Blackwell Science, 1998: pp. 319–332.

[142]   Paulson OB, Strandgaard S, Edvinsson L. Cerebral autoregulation. *Cerebrovasc Brain Metab Rev.* 1990; 2: pp. 161–192.

[143]   Euser AG, Cipolla MJ. Cerebral blood flow autoregulation and edema formation during pregnancy in anesthetized rats. *Hypertension.* 2007; 49: pp. 334–340.

[144]   Lassen NA, Agnoli A. Upper limit of autoregulation of cerebral blood flow: on the pathogenesis of hypertensive encephalopathy. *Scand J Clin Lab Invest.* 1972; 30: pp. 113–115.

[145]   Johansson B, Li C-L, Olsson Y, Klatzo I. Effect of acute arterial hypertension on the blood–brain barrier to protein tracers. *Acta Neuropathol.* 1970; 16: pp. 117–124.

[146]   Kontos HA, Wei EP, Navari RM, Levasseur JE, Rosenblum WI, Patterson JL, Jr. Responses of cerebral arteries and arterioles to acute hypotension and hypertension. *Am J Physiol.* 1978; 234: pp. H371–H383.

[147]   Cipolla MJ. Brief review: Cerebrovascular function during pregnancy and eclampsia. *Hypertension.* 2007; 50(1): pp. 14–24.

[148]   Phillips SJ, Whisnant JP. Hypertension and the brain. *Arch Intern Med.* 1992; 152: pp. 938–945.

[149]   Strandgaard S, Paulson OB. Hypertensive disease and the cerebral circulation. In: *Hypertension: Pathophysiology, Diagnosis, and Management*, Laragh JH, Brenner BM (Eds.). New York: Raven Press, 1990: pp. 399–416.

[150]   Strandgaard S, MacKenzie ET, Sengupta D, Rowan JP, Lassen NA, Harper AM. Upper limit of autoregulation of cerebral blood flow in the baboon. *Circ Res.* 1974; 34: pp. 434–440.

[151]   Byrom FB. The pathogenesis of hypertensive encephalopathy and its relation to the malignant phase of hypertension. *Lancet.* 1954; 2: pp. 201–211.

[152]   Meyer JS, Waltz AG, Gotoh F. Pathogenesis of cerebral vasospasm in hypertensive en-

cephalopathy: II. Nature of increased irritability of smooth muscle of pial arterioles in renal hypertension. *Neurology*. 1960; 10: pp. 859–867.

[153] Byrom FB. *The Hypertensive Vascular Crisis: An Experimental Study*. London, Heinemann, 1969.

[154] Giese J. Acute hypertensive vascular disease. II. Studies on vascular reaction patterns and permeability changes by means of vital microscopy and colloidal tracer technique. *Acta Pathol Microbiol Scand*. 1964; 62: pp. 497–515.

[155] Skinhøj E, Strandgaard S. Pathogenesis of hypertensive encephalopathy. *Lancet*. 1973; 1: pp. 461–462.

[156] Tamaki K, Sadoshima S, Baumbach GL, Iadecola C, Reis DJ, Heistad DD. Evidence that disruption of the blood–brain barrier precedes reduction in cerebral blood flow in hypertensive encephalopathy. *Hypertension*. 1984; 6(Suppl I): pp. I75–I81.

[157] Johnson, PC. The myogenic response. In: *Handbook of Physiology. The Cardiovascular System. Vascular Smooth Muscle*, Section 2, Vol. II. Bethesda, MD: American Physiological Society, 1981: pp. 409–442.

[158] Heistad DD, Marcus ML, Abboud FM. Role of large arteries in regulation of cerebral blood flow in dogs. *J Clin Invest*. 1978; 62: pp. 761–768.

[159] Murphy S, Rich G, Orgren KI, Moore SA, Faraci FM. Astrocyte-derived lipoxygenase product evokes endothelium-dependent relaxation of the basilar artery. *J Neurosci Res*. 1994; 38: pp. 314–318.

[160] Filosa JA, Bonev AD, and Nelson MT. Calcium dynamics in cortical astrocytes and arterioles during neurovascular coupling. *Circ Res*. 2004; 95: pp. e73–e81.

[161] Mulligan SJ, MacVicar BA. Calcium transients in astrocyte endfeet cause cerebrovascular constrictions. *Nature*. 2004; 431: pp. 195–199.

[162] Zonta M, Angulo MC, Gobbo S, Rosengarten B, Hossmann KA, Pozzan T, and Carmignoto G. Neuron-to-astrocyte signaling is central to the dynamic control of brain microcirculation. *Nat Neurosci*. 2003; 6: pp. 43–50.

[163] Masamoto K, Tanishita K. Oxygen transport in brain tissue. *J Biomech Eng*. 2009; 131: pp. 74–82.

[164] Steiner LA AJ, Gupta AK, Menon DK. Cerebral oxygen vasoreactivity and cerebral tissue oxygen reactivity. *Br J Anaesth*. 2003; 90: pp. 774–786.

[165] Taguchi H, Heistad DD, Kitazono T, Faraci FM. ATP-sensitive K+ channels mediate dilatation of cerebral arterioles during hypoxia. *Circ Res*. 1994; 74: pp. 1005–1008.

[166] Golanov EV, Reis DJ. Oxygen and cerebral blood flow. In: *Primer on Cerebrovascular Diseases*, Welch KMA, Caplan LR, Reis DJ, Siesjo BK, Weir B (Eds.). San Diego, CA: Academic Press, 1997.

[167] Reivich M. Arterial PCO2 and cerebral hemodynamics. *Am J Physiol*. 1964; 206: pp. 25–35.

[168]   Kety SS, Schmidt CF. The effects of altered arterial tensions of carbon dioxide and oxygen on cerebral blood flow and cerebral oxygen consumption of normal young men. *J Clin Invest.* 1948; 27: pp. 484–492.

[169]   Kontos HA, Raper AJ, Patterson JL. Analysis of vasoactivity of local pH, PCO2 and bicarbonate on pial vessels. *Stroke.* 1977; 8: pp. 358–360.

[170]   Kontos HA, Wei EP, Raper AJ, Patterson JL Jr. Local mechanism of CO2 action of cat pial arterioles. *Stroke.* 1977; 8: pp. 226–229.

[171]   Iadecola C. Does nitric oxide mediate the increases in cerebral blood flow elicited by hypercapnia? *Proc Natl Acad Sci USA.* 1992; 89: pp. 3913–3916.

[172]   Pickard JD, Mackenzie ET. Inhibition of prostaglandin synthesis and the response of baboon cerebral circulation to carbon dioxide. *Nat New Biol.* 1973; 245: pp. 187–188.

[173]   Ehrlich P. Das Sauerstoff-Bedürfnis des Organismus. Eine farbenanalytische Studie. PhD thesis. Berlin: Herschwald, 1885.

[174]   Goldmann E. *Vitalfärbung am Zentralnervensystem. Beitrag zur Physio-Pathologie des Plexus chorioideus und der Hirnhäute.* Berlin: Abh Königl Preuss Akad Wiss, 1913; 1: pp. 1–61.

[175]   Choi YK, Kim KW. Blood-neural barrier: its diversity and coordinated cell-to-cell communication. *BMB Rep.* 2008; 41: pp. 345–352.

[176]   Saunders NR, Ek CJ, Habgood MD, Dziegielewska KM. Barriers in the brain: a renaissance? *Trends Neurosci.* 2008; 31: pp. 279–286.

[177]   Wolburg H, Noell S, Mack A, Wolburg-Buchholz K, Fallier-Becker P. Brain endothelial cells and the glio-vascular complex. *Cell Tissue Res.* 2009; 335: pp. 75–96.

[178]   Kimelberg HK. Water homeostasis in the brain: basic concepts. *Neuroscience.* 2004; 129(4): pp. 851–860.

[179]   Skipor J, Thiery JC. The choroid plexus–cerebrospinal fluid system: undervaluated pathway of neuroendocrine signaling into the brain. *Acta Neurobiol Exp (Wars).* 2008; 68(3): pp. 414–428.

[180]   Czosnyka M, Czosnyka Z, Momjian S, Pickard JD. Cerebrospinal fluid dynamics. *Physiol Meas.* 2004; 25: pp. R51–R76.

[181]   Proescholdt MG, Hutto B, Brady LS, Herkenham M. Studies of cerebrospinal fluid flow and penetration into brain following lateral ventricle and cisterna magna injections of the tracer [14C]inulin in rat. *Neuroscience.* 2000; 95: pp. 577–592.

[182]   Betz AL, Goldstein GW, Katzman R. Blood–brain–cerebrospinal fluid barriers. In: *Basic Neurochemistry: Molecular, Cellular and Medical Aspects*, Seigel GJ (Ed.). New York: Raven Press, 1994: pp. 681–702.

[183]   Faraci FM, Mayhan WG, Williams JK, Heistad DD. Effects of vasoactive stimuli on blood flow to choroid plexus. *Am J Physiol.* 1988; 254(2 Pt 2): pp. H286–H291.

[184]   Nilsson C, Stahlberg F, Thomsen C, Henriksen O, Hering M, Owman C. Circadian varia-
tion in human cerebrospinal fluid production measured by magnetic resonance imaging. *Am
J Physiol.* 1992; 262: pp. R20–R24.

[185]   Reese TS, Karnovsky MJ. Fine structural localization of a blood brain barrier to exogenous
peroxidase. *J Cell Biol.* 1967; 34: pp. 207–217.

[186]   Brightman MW, Reese TS. Junctions between intimately apposed cell membranes in the
vertebrate brain. *J Cell Biol.* 1969; 40: pp. 648–677.

[187]   Deane R, Zlokovic BV. Role of blood brain barrier in the pathogenesis of Alzheimer's disease. *Curr Alzheimer Res.* 2007; 4: pp. 191–197.

[188]   Ueno M. Molecular anatomy of the brain endothelial barrier: an overview of the distribu-
tional features. *Curr Med Chem.* 2007; 14: pp. 1199–1206.

[189]   Duvernoy HM, Risold PY. The circumventricular organs: an atlas of comparative anatomy
and vascularization. *Brain Res Rev.* 2007; 56: pp. 119–147.

[190]   Brightman MW, Tao-Cheng JH. In: *The Blood–Brain Barrier,* Pardridge WM (Ed.). New
York: Raven Press, 1993: pp. 107–125.

[191]   Kniesel U, Wolburg H. Tight junctions of the blood–brain barrier. *Cell Mol Neurobiol.* 2000;
20: pp. 57–76.

[192]   Gumbiner B, Lowenkopf T, Apatira D. Identification of a 160-kDa polypeptide that binds
to the tight junction protein ZO-1. *Proc Natl Acad Sci USA.* 1991; 179: pp. 3460–3464.

[193]   Haskins J, Gu L, Wiichen ES, Hibbard J, Stevenson BR. ZO-3, a novel member of the
MAGUK protein family found at the tight junction, interacts with ZO-1 and occludin.
J Cell Biol. 1998; 141: pp. 199–208.

[194]   Furuse M, Sasaki H, Tsukita S. Manner of interaction of heterogeneous claudin species
within and between tight junction strands. *J Cell Biol.* 1999; 147: pp. 891–903.

[195]   Morita K, Sasaki H, Fujimoto K, Furuse M, Tsukita S. Claudin-11/OSP-based tight junctions of myelin sheaths in brain and Sertoli cells in testis. *J Cell Biol.* 1999; 145: pp. 579– 588.

[196]   Citi S, Sabanay H, Jakes R, Geiger B, Kendrick-Jones J. Cingulin, a new peripheral component of tight junctions. *Nature.* 1988; 333: pp. 272–276.

[197]   Stevenson BR, Heintzelman MB, Anderson JM, Citi S, Mooseker MS. ZO-1 and cingulin: tight junction proteins with distinct identities and localizations. *Am J Physiol.* 1989; 257:
pp. C621–C628.

[198]   Furuse M, Hirase T, Itoh M, Nagafuchi A, Yonemura S, Tsukita S. Occludin: a novel integral membrane protein localizing at tight junctions. *J Cell Biol.* 1993; 123: pp. 1777–1788.

[199]   Ando-Akatsuka Y, Saitou M, Hirase T, Kishi M, Sakakibara A, Itoh M, Yonemura S,
Furuse M, Tsukita S. Interspecies diversity of the occludin sequence: cDNA cloning of
human, mouse, dog, and rat-kangaroo homologues. *J Cell Biol.* 1996; 133: pp. 43–47.

[200]   Mitic LL, Van Itallie CM, Anderson JM. Molecular physiology and pathophysiology of tight junctions I. Tight junction structure and function: lessons from mutant animals and proteins. *Am J Physiol.* 2000; 279: pp. G25–G254.

[201]   Ebnet K, Schulz CU, Meyer Zu Brickwedde MK, Pendl GG, Vestweber D. Junctional adhesion molecule interacts with the PDZ domain-containing proteins AF-6 and ZO-1. *J Biol Chem.* 2000; 275: pp. 27979–27988.

[202]   Aurrand-Lions M, Johnson-Leger C, Wong C, Du Pasquier L, Imhof BA. Heterogeneity of endothelial junctions is reflected by differential expression and specific subcellular localization of the three JAM family members. *Blood.* 2001; 98: pp. 3699–3707.

[203]   Watabe M, Nagafuchi A, Tsukita S, Takeichi MJ. Induction of polarized cell–cell association and retardation of growth by activation of the E-cadherin–catenin adhesion system in a dispersed carcinoma line. *Cell Biol.* 1994; 127: pp. 247–256.

[204]   Lampugnani MG, Corada M, Caveda L, Breviario F, Ayalon O, Geiger B, Dejana E. The molecular organization of endothelial cell to cell junctions: differential association of plako-globin, beta-catenin, and alpha-catenin with vascular endothelial cadherin (VE-cadherin). *J Cell Biol.* 1995; 129: pp. 203–217.

[205]   Matter K, Balda MS. Signalling to and from tight junctions. *Nat Rev Mol Cell Biol.* 2003; 4: pp. 225–236.

[206]   Cipolla MJ. Stroke and the blood–brain interface. In: *Blood–Brain Barrier Interfaces*, Spray D, Dermietzel R (Eds.). New York: Wiley, 2006.

[207]   Geockeler ZM, Wysolmerski RB. Myosin light chain kinase-regulated endothelial cell contraction: relationship between isometric tension, actin polymerization and myosin phosphorylation. *J Cell Biol.* 1995; 130: pp. 613–627.

[208]   Lum H, Malik AB. Regulation of vascular endothelial barrier function. *Am J Physiol.* 1994; 267: pp. L223–L241.

[209]   Yuan SY. Signal transduction pathways in enhanced microvascular permeability. *Microcirculation.* 2000; 7: pp. 395–405.

[210]   Garcia JG, Davis HW, Patterson CE. Regulation of endothelial cell gap formation and barrier dysfunction: role of myosin light chain phosphorylation. *J Cell Physiol.* 1995; 163: pp. 510–522.

[211]   Pardridge WM. Blood–brain barrier delivery. *Drug Discov Today.* 2007; 12: pp. 54–61.

[212]   Broadwell RD, Balin BJ, Salcman M. Transcytotic pathway for blood-borne protein through the blood–brain barrier. *Proc Natl Acad Sci USA.* 1988; 85: pp. 632–636.

[213]   Broadwell RD. Pathways into, through, and around the fluid–brain barriers. *NIDA Res Monogr.* 1992; 120: pp. 230–258.

[214]   Berne RM, Levy MN. The microcirculation and lymphatics. In: *Physiology*, Berne RM, Levy MN, Koeppen BM, Stanton BA (Eds.). St. Louis, MO: Mosby, 1998: pp. 429–441.

[215]   Rubin LL, Staddon JM. The cell biology of the blood–brain barrier. *Annu Rev Neurosci.* 1999; 22: pp. 11–28.

[216]   Kniesel U, Wolburg H. Tight junctions of the blood–brain barrier. *Cell Mol Neurobiol.* 2000; 20: pp. 57–76.

[217]   Baldwin A, Wilson L. Endothelium increases medial hydraulic conductance of aorta, possibly by release of ERDF. *Am J Physiol.* 1993; 264: pp. H26–H32.

[218]   Baldwin AL, Wilson LM, Gradus-Pizlo I, Wilensky R, and March K. Effect of atherosclerosis on transmural convection and arterial ultrastructure: implications for local intravascular drug delivery. *Arterioscler Thromb Vasc Biol.* 1997; 17: pp. 3365–3375.

[219]   Shou Y, Jan KM, Rumschitzki DS. Transport in rat vessel walls. I. Hydraulic conductivities of the aorta, pulmonary artery, and inferior vena cava with intact and denuded endothelia. *Am J Physiol.* 2006; 291: pp. H2758–H2771.

[220]   Davson H, Oldendorf WH. Transport in the central nervous system. *Proc R Soc Med.* 1967; 60: pp. 326–328.

[221]   Reinhart CA, Gloor SM. Co-culture blood–brain barrier models and their use for pharmatoxicologic screening. *Toxicol Vitro.* 1997; 11: pp. 513–518.

[222]   Bauer HC, Bauer H. Neural induction of the blood–brain barrier: still an enigma. *Cell Mol Neurobiol.* 2000; 20: pp. 13–28.

[223]   Quick AM, Cipolla MJ. Pregnancy-induced upregulation of aquaporin 4 in brain and its role in eclampsia. *FASEB J.* 2004; 19: pp. 170–175.

[224]   Price DL, Ludwig JW, Mi H, Schwarz TL, Ellisman MH. Distribution of rSlo Ca2+-activated K+ channels in rat astrocyte perivascular endfeet. *Brain Res.* 2002; 956: pp. 183–193.

[225]   Klatzo I. Neuropathologic aspects of brain edema. *J Neuropathol Exp Neurol.* 1967; 26: pp. 1–14.

[226]   Dodson RF, Chu LW, Welch KM, Achar VS. Acute tissue response to cerebral ischemia in the gerbil: an ultrastructural study. *J Neurol Sci.* 1977; 33: pp. 161–170.

[227]   Chen Y, Swanson RA. Astrocytes and brain injury. *J Cereb Blood Flow Metab.* 2003; 23(2): pp. 137–149.

[228]   Anderson CM, Swanson RA. Astrocyte glutamate transport: review of properties, regulation and physiological functions. *Glia.* 2000; 32: pp. 1–14.

[229]   Walz W, Hertz L. Functional interactions between neurons and astrocytes. Part II: potassium homeostasis at the cellular level. *Prog Neurobiol.* 1983; 20: pp. 133–183.

[230]   Tanaka J, Toku K, Zhang B, Ishihara K, Sakanaka M, Maeda N. Astrocytes prevent neuronal death induced by reactive oxygen and nitrogen species. *Glia.* 1999; 28: pp. 85–96.

[231]   Fenstermacher JD. Volume regulation of the central nervous system. In: *Edema.* Staub NC, Taylor AE (Eds.). New York: Raven Press, 1984: pp. 383–404.